A^3 & HIS ALGEBRA

A^3 & HIS ALGEBRA

How a Boy from Chicago's West Side Became a Force in American Mathematics

Nancy E. Albert

iUniverse, Inc.
New York Lincoln Shanghai

A^3 & HIS ALGEBRA
How a Boy from Chicago's West Side Became a Force in American Mathematics

Copyright © 2005 by Nancy E. Albert

All rights reserved. No part of this book may be used or reproduced by any means, graphic, electronic, or mechanical, including photocopying, recording, taping or by any information storage retrieval system without the written permission of the publisher except in the case of brief quotations embodied in critical articles and reviews.

iUniverse books may be ordered through booksellers or by contacting:

iUniverse
2021 Pine Lake Road, Suite 100
Lincoln, NE 68512
www.iuniverse.com
1-800-Authors (1-800-288-4677)

The author and publisher gratefully acknowledge permission to reprint material from the following: University of Chicago Special Collections; Archives of the American Mathematical Society; University of Göttingen Department of Manuscripts and Rare Books; Princeton University archives; Institute for Advanced Study archives; Peter Roquette, "The Brauer-Hasse-Noether Theorem in Historical Perspective," http://www.roquette.uni-hd.de; H. P. Petersson, "Albert Algebras: Their Meaning for Present-day Mathematics" (unpub.); J. J. O'Connor and E. F. Robertson, MacTutor History of Mathematics Archive, University of St. Andrews, Scotland; Nathan Jacobson, "Abraham Adrian Albert, 1905–1972," *Bull. Amer. Math. Soc.* 80 (1974), 1075-1100, reprinted in *A. Adrian Albert Collected Mathematical Papers*, (2 vols.), Providence: AMS, 1993; Irving Kaplansky, "Abraham Adrian Albert, a Biographical Memoir," *Biographical Memoirs*, National Academy of Sciences 51 (1980), 3-22; Daniel Zelinsky, "A. A. Albert," American Mathematical Monthly 80 (1973), 661-665; Title page photo: H. Anne Plettinger.

ISBN-13: 978-0-595-32817-8
ISBN-10: 0-595-32817-2

Printed in the United States of America

For Sandra, Debra,
Jo Jo, Roy,
and Peter

Contents

Preface ix
Author's Note and Acknowledgments xi
Prologue 1

PART ONE: BEGINNINGS

1 Roots 7
2 Birth of a University in a New City 15
3 The First Two Generations at the Mathematics Department 25
4 The Early Years at Chicago and Princeton 34
5 Kinships and Kindred Spirits 49

PART TWO: FRAMEWORK FOR A MATHEMATICAL MIND

6 Mathematese and Mathemagicians 57
7 From Al Jabr to Abstract Algebra 63
8 Flowering of Abstract Algebra over a Shifting Field 72

PART THREE: YEARS OF DISCOVERY

9 Explorers in Vector Space 79
10 The Proof 85
11 Traveling Riemann's Universe 104
12 A New School of Algebra 114

PART FOUR: BRANCHING OUT IN MIDLIFE

13   Mathematics during World War II 125
14   South of the Border 142
15   Hyde Park in the Forties and Beyond 152
16   Albert Algebras 159
17   Defending His Country 163
18   The University's Solution to Urban Decay 173
19   East of the Tracks 177

PART FIVE: ELDER STATESMAN

20   Years of Triumph and Tragedy 185
21   Urban Removal and Condomania 203
22   E. H. Moore at the Entry 207
23   An Entirely New Dimension 213
24   Turmoil on Campus 223
25   Education Guru 229
26   Building Bridges across the Globe 235
27   Diplomacy for Mathematics 247
28   Mathematical Children 253

Appendix: Doctoral Dissertations Supervised by A. A. Albert 267
Notes 271
Select Bibliography 321
Index 327
About the Author 351

Preface

In 1951, after completing a Ph.D. in mathematics at the University of Wisconsin, I applied for ten academic positions. I was rejected for nine of them. Thanks to my thesis, excellent recommendations, and A. A. Albert, I spent my next two extremely productive years as Research Instructor in the Mathematics Department at the University of Chicago. Times were tough and I was fortunate. The Depression and the aftermath of World War II were still having their repercussions on the mathematical community in the United States.

In 1952, while browsing through used bookstores near the University of Chicago during my lunch hours, I became acutely aware of how rapidly mathematics in the United States had progressed, from far behind European mathematics at the turn of the century to well ahead and at the leading edge by the middle of the twentieth century.

A. A. Albert loved mathematics and the mathematicians who worked at the frontiers of knowledge. He was convinced that through their work they could contribute a lot to the betterment of society. He played no small part in seeing to it that promising research mathematicians could carry out this mission.

In a very readable account, his daughter Nancy Albert has brought A. A. Albert and his era back to life. Recounting Albert's career, from his humble beginnings on Chicago's Maxwell Street to becoming Chairman of Chicago's Mathematics Department, then Dean of Chicago's Physical Sciences Division, to his numerous honors and important contributions to the national defense, she recreates the excitement of the times—a time when both government and academia were pushing mathematical research and much was happening that was good for mathematics.

Reno, Nevada   Erwin Kleinfeld

Author's Note and Acknowledgments

This is the true story of one man's struggle to follow his passion. A first generation American, he began his journey in Chicago's impoverished Maxwell Street neighborhood. Lacking financial means, he turned to his only assets: a powerful intellect, a loving father who fostered it, and an unsinkable determination. These gifts propelled the budding scholar over the obstacles of poverty, prejudice, and personal tragedy to reach his dream—becoming a mathematician.

Like the "can-do" city he inhabited, the young man pulled himself up by his bootstraps, determined to succeed. In the face of a family crisis that threatened his ability to stay in school, he fast-tracked his education to finish in record time. Once he reached that milestone, there was no stopping him.

As a researcher of pure mathematics, he made pioneering advances in modern higher algebra. As an applied mathematician, he played a major role on the national, and, eventually, the world stage. He lived through two world wars and two major conflicts—Korea and Vietnam. In all but the first of these, he defended his country behind the scenes, often behind the velvet curtains of the United States National Security Agency.

When the young man entered his chosen field of mathematics, he found it to be a neglected science in America. The Great Depression exacerbated the situation, leaving too few Ph.D. mathematicians chasing still fewer jobs in the field. He took it as his mission to help change the attitudes of American policymakers toward mathematics. He served in the vanguard of those who proved, in World War II and thereafter, that mathematics was not a "dead science," but a vital instrument for the security of the country. By the time he left his desk, mathematics was a thriving field contributing to the nation's well-being in ever increasing ways.

During his forty-year career, he traveled the globe as one of the leaders of the world's mathematical community. Despite his travels, he remained steadfast in his allegiance to his alma mater, the University of Chicago, where he eventually became Chairman of the Mathematics Department and later, Dean of Physical Sciences.

The man who made that journey is my late father, Abraham Adrian Albert. This memoir offers a personal and revealing look at the multidimensional life of an academic who had a lasting impact on his profession.

It is intended for mathematical and general audiences alike. Some background information is helpful for an understanding of A.A. Albert's mathematics. For readers who are new to algebra, Chapter 6 cuts through some of the mystique of mathematics by taking a whimsical approach to defining basic terms and concepts. Chapters 7 and 8 provide a frame of reference for my father's work by laying out the developments that led up to it.

Chapters 9–12 and 16 describe the mathematical travails encountered by A.A. Albert, his mentors, and his contemporaries in their quest to unravel the seemingly inscrutable secrets of algebra. The Notes at the end of this book provide citations to their published work.

Throughout the narrative, I have attempted to demonstrate the intimate relationship between the work of algebraists and applications in our everyday lives. Chapters 13 and 17 delve into the ways that A.A. Albert and others used applied algebra in their national defense work.

A^3 *and His Algebra* puts a human face on the often off-putting discipline of algebra. I hope my father's story will help make the subject more accessible to readers who would normally shy away from it.

Adrian, as he preferred to be called, had many friends and colleagues who helped to tell his story. After his death, several of his closest colleagues wrote separate biographical sketches that were invaluable in preparing this book. They were: I. N. Herstein, Nathan Jacobson, Irving Kaplansky, and Daniel Zelinsky. These memoirs were later republished by the American Mathematical Society along with many of Adrian's papers in the two-volume set, *A. Adrian Albert, Collected Mathematical Papers*. Thanks are owed to the editors: Richard E. Block, Nathan Jacobson, J. Marshall Osborn, David J. Saltman, and Daniel Zelinsky.

My greatest debt in preparing this memoir is to my mother. Frieda Albert was not only an important helpmate to Adrian while he pursued his mathematical career, but a careful steward of his papers and letters as well. I also want to thank my uncle, Dan Davis, who kept me supplied with family data.

There are many others who helped to make this biography possible. I am deeply indebted to the scientists who provided me with information, some of whom patiently guided me through the intricate nuances of algebra. They include: Irving Kaplansky, Erwin Kleinfeld, Margaret Kleinfeld, Louis Rowen, Holger Petersson, Peter Roquette, David Saltman, Stephen Stigler, David Oxtoby, Norman Lebovitz, William Reid, Paul R. Halmos, and Kevin Corlette.

I am extremely grateful to Daniel Meyer, Alice Schreyer, and Eileen Ielmini at the University of Chicago libraries. Many of the letters and other materials used in this biography are housed in the A. Adrian Albert Papers in Special Collections at the University's Joseph Regenstein Library. My thanks also to Timothy Engels and the wonderful staff at the John Hay Library of Brown University for their kind assistance in accessing the archives of the American Mathematical Society and to Christine M. Thivierge at the American Mathematical Society.

Thanks are owed to the American Mathematical Society, the Institute for Advanced Study, Princeton University, and the University of Göttingen for permission to use their archival material. I would like to thank Dan Sharon, Senior Reference Librarian at the Asher Library of the Spertus Institute of Jewish Studies. The author is also grateful for assistance by the Princeton Historical Society, the Hyde Park Historical Society, and Kevin Leonard at the Northwestern University Library archives.

This biography drew on a fount of valuable information from the MacTutor History of Mathematics Archive of the University of St. Andrews in Scotland, which is available on the Internet at http://www-groups.dcs.st-andrews.ac.uk/. I am indebted to J. J. O'Connor, E. F. Robertson, and others who compiled biographies of many famous mathematicians.

"A Cubed," as A.A. Albert was dubbed by students and colleagues, left behind a reservoir of good will among those whose lives he touched. I attribute the generous cooperation of his former students to their recollections of him. They include Richard E. Block, Robert B. Brown, Harley Flanders, Marguerite Straus Frank, Murray Gerstenhaber, Lou Kahn, Benjamin Muckenhoupt, Robert Oehmke, J.

Marshall Osborn, Eugene C. Paige, Reuben Sandler, Richard D. Schafer, John G. Thompson, and Daniel Zelinsky.

My deepest thanks go to the mathematicians, Daniel Zelinsky, George Glauberman, and Gerald L. Alexanderson, who took the time to review my manuscript. Their insights were indispensable. They should not be held accountable for any errors that I may have made, as I did not always heed their advice. I bear the sole responsibility for them. Finally, heartfelt thanks to Jerry Jordan and Marshall Hartman for their thoughtful comments, and a special thanks to Zelda Zelinsky for providing a great deal of food for thought.

Chicago, Illinois *Nancy E. Albert*

Prologue

The nickname "A Cubed" suited him. Father was as multidimensional as the objects he investigated.

Around the University, most people called him "Albert." To his close friends, he was "Adrian." At home, Mother simply addressed him as "A." She never explained which of his three names the A stood for.

The initial matched his type A personality. He constantly drove himself to excel. But you wouldn't know it. A mild-mannered man, he had the relaxed air of one whose accomplishments came easily.

Father was a product of Chicago. And then again, many of the ingredients came from elsewhere. His roots influenced him in ways I could not fathom. The intermittent periods he spent in America's Northeast and Southwest provided much of the raw material.

He was left-handed. Except when he wrote with his right hand.

His peers saw him as a pure mathematician. Yet, unlike most American mathematicians, he straddled the fence between pure and applied mathematics.

He had an innocent, boyish quality. On the other hand, he treated sensitive national security issues with the utmost gravity.

He was a shy, reticent person. But when he spoke, he could persuade government bureaucrats to reverse their thinking about the strategic value of pure mathematics.

Our home itself was something of a paradox. An off-campus rental apartment in Chicago's academic enclave of Hyde Park, its simple furnishings reflected the lifestyle of an underpaid college professor. A maroon faux Oriental rug set the tone of the living room where Father scribbled equations at a small writing table, sometimes after a sudden burst of inspiration at three in the morning. A modest corner bookcase housed a variety of literature, including Father's texts. A nook in the foyer offered just enough space for a two-tiered telephone table, handmade by Father himself.

In that understated dwelling, Mother threw elaborate dinner parties for the entire Mathematics Department embellished by fine china, crystal goblets, and a bedazzling array of silver-plated cutlery. A typical Hyde Park apartment, its living and dining rooms sat on opposite ends, like two ships across a divide. The dining room merited the most expensive of our furniture pieces—a massive claw-footed hardwood table and matching buffet that hosted many mathematical gatherings.

The Hyde Park neighborhood we inhabited was diverse and multifaceted as well. Professors, students, doctors, businessmen, and factory workers called it home. Our building at the corner of 57th Street and Drexel Boulevard reflected the mix of the community. Historian Daniel Boorstin and his family lived in the apartment directly below ours, opposite that of a steelworker.

The University of Chicago campus dominated the neighborhood and the people in it. Growing up in Hyde Park, the campus seemed to me a magical place. Lofty ideals appeared to rise from the spires and pointed arches of neo-gothic, ivy-covered halls that bore the aspect of medieval cathedrals. Carillon concerts resounded across the green from the giant bell tower in Rockefeller Chapel. On our way home from school, gargoyles atop the University's Hull Gate beckoned my classmates and me into the Quadrangles to admire the lily pads and goldfish in Botany Pond.

Across the street from Hull Gate loomed the seemingly innocent bleachers of the University's former football stadium, Stagg Field. My friends and I had no idea that, beneath the stands, Manhattan Project scientists had developed the world's first controlled nuclear chain reaction. Nor were we aware that, during the war, the Manhattan Project occupied the mathematics building, Eckhart Hall, as well as other buildings on campus.

My parents had a close relationship. But that was not apparent to us—their generation avoided displays of affection in front of the children. They appeared cool and detached. Father's love letters to Mother told a different story.

By the time I arrived, the youngest of three children, Father had developed a demanding set of responsibilities. The University had grown to be as much a part of his family as his genetic kin. His travel to meetings and consultancies was consuming. It was not easy to get to know him.

He was a reserved, private person. His driving force, his background, his strengths and weaknesses, the nature of his work, were an enigma to me. I knew little about the arrow-like progression of his career or how he achieved it.

He rarely spoke about his work. Besides being difficult to comprehend, so much of it was meant to be kept a secret. For example, I always believed that our lengthy stays in California were designed to allow Mother to visit her family. I had no clue as to his other mission in the Golden State.

It has taken me many years to unravel the incongruities of my father's life and of the University whose existence was so intimately intertwined with his own. In order to search for the roots of these disparities, I had to go back to the very beginnings.

PART ONE

Beginnings

Chapter 1

Roots

Elias (Elijah) Albert grew up in the eastern European city of Vilna, the capital of Lithuania. The history of Lithuania dates back to the 13th century when residents of the region banded together to resist a group of Teutonic invaders.

By the beginning of the 17th century, Vilna was a preeminent center for rabbinical studies.[1] During the second half of the 18th century, a "wise man" named Elijah ben Solomon became the leader of Vilna's Jewry.

Family legend has it that Elias Albert was descended from Elijah ben Solomon, also known as the "Gaon (Genius) of Vilna." According to family lore, Elias was a fifth generation descendant of that sage, who lived from 1728 to 1797.[2] Himself descended from a long line of scholars, the Vilna Gaon became famous in early childhood for prodigious feats of rapid learning.[3]

Elijah ben Solomon vigorously opposed mysticism. He devoted himself to incessant learning, reportedly sleeping no more than two hours a night.[4] Elijah did not restrict his studies to religious texts, since he believed that one must be well versed in secular knowledge to fully understand the scriptures.[5] Among the subjects he studied were astronomy, geometry, algebra, and geography.[6]

The Vilna Gaon spent the first forty years of his life studying, and the remainder, imparting what he had learned to his pupils.[7] Although he refused to accept the position of Rabbi, he became the spiritual head not only of Vilna, but also of the entire Lithuanian and Russian Jewry. He fulfilled the role of "rabbi" in the sense that rabbi meant "teacher."

He attracted a large number of disciples, and his influence had a lasting impact on Judaism.[8]

After Rabbi Elijah's death, his pupils published 40 volumes of his teachings on a variety of subjects. Two of Elijah's sons also worked extensively on publishing his writings and edited a number of Elijah's manuscripts on algebra and trigonometry.[9]

Elijah ben Solomon's brother was named Avraham and one of his sons was Avraham. The names Elijah and Avraham run throughout the generations of the Vilna Gaon's descendants.

Lithuania was defeated in battle at the end of the 18th century and became part of Russia in 1795. The country remained under Russian control for over a century afterwards. By the time of Elias Albert's birth in 1854, Lithuania belonged to Russia.

Elias was the elder of two brothers. He fled Vilna at the age of 14, purportedly over some sort of rift in his family. By leaving when he did, the youth also avoided conscription into the Russian army.

Afterwards, he lived in England for about a dozen years. He adopted "Albert" as his surname, after Queen Victoria's prince consort. Elias went to great lengths to ensure that his original surname remained a mystery.

In England, he acquired an education and taught school. He learned to write in a beautiful, Spenserian script, to love good literature, to write poetry, and to eat and live like an Englishman.[10]

Elias arrived on American soil in 1880. He and a partner attempted to earn a living by starting up a picture frame company in Hartford, Connecticut. Meanwhile, he wrote articles for a Yiddish language newspaper and indulged his love of literature by collecting first edition books.[11]

In his late twenties, he moved to Chicago and married a woman who, like himself, came from Vilna. The marriage ended tragically when his wife perished in childbirth along with their only child.

Fannie Fradkin was born in Kiev, Russia in 1869. She arrived in Chicago with her older married sister when she was only thirteen. By the time she met Elias, Fannie was a widow, struggling to support herself and her small children. She was a plain woman, but industrious. She ran a small retail fish market in a rented storefront.

Elias had waited a decade to remarry. He had two major reasons for marrying Fannie—her two adorable children. His union with Fannie fulfilled his longing for fatherhood. Her fish market, on the other hand, was not his cup of tea. It took some doing for him to adjust his English sensibilities to its pungent aroma.

The low-margin, high-maintenance business barely made ends meet. Still, it was a living. Together, they could make it viable.

Upon marrying Fannie, his fortunes multiplied. Not only did they have Fannie's two children, the couple had three children together. They greeted their first child in 1901—a little girl they named Minnie. In Hebrew, her name was "Minucha," meaning "rest."

Like thousands of Eastern European Jews who poured into Chicago during the late nineteenth and early twentieth centuries, they lived on Chicago's Near West Side, which became known as the "Poor Jews Quarter" and somewhat later as the Maxwell Street area.[12] The predominantly Jewish area just southwest of the central business district was one of the poorest parts of the city. Residents made do with ramshackle dwellings where rent was cheap. Sweatshops and pushcarts abounded in the neighborhood.

A baby boy arrived on November 9, 1905. Elias was overjoyed. The 51-year old man finally had his first son.

They named him Abraham, after Elias' father. In his youth, the boy was "Abraham" or "Abe." Like other immigrants' sons, he soon recognized that assimilation was the path to success. As he approached adulthood, he dropped the ethnic-sounding "Abraham," adopting instead his English-sounding middle name "Adrian."

Elias was a loving father and stepfather. Adrian's half-sister Esther, 12 years his senior, recalled that Elias took her to the opera. She looked up to the kindly man.

As a young child, Adrian had blue eyes and golden blond hair, like his father. Esther doted on her baby brother and cried when his long, golden curls were shorn at the age of five. She considered him brilliant and dubbed the studious child "Little Professor."

In his youth, he developed a facility with numbers. The precocious youngster could calculate numbers in his head at lightning speed. Elias recognized his son's potential and did his utmost to shield the budding scholar from mundane chores. "Let the boy study," he would say.

At the age of 11, Esther dropped out of school to help Fannie and Elias run the family market.[13] It was fairly commonplace for youngsters of that era to begin working at an early age. But for his father's intervention, Adrian, too, would have been obliged to quit school.

The philosophy in public schools when he was growing up favored forcing left-handed students to learn to write with their right hand. Some students suffered as a result of the practice.[14] Adrian simply learned to write with both hands and used the skill to his advantage.

While he was still a boy, Esther married and moved to Iron Mountain, Michigan. Elias and Fannie temporarily moved the entire family to Iron Mountain, and Adrian attended grade school there from 1914 to 1916.

After two years in Michigan, Elias brought his family back to Chicago. Instead of returning to the old Maxwell Street neighborhood, they moved into a new Jewish neighborhood called Lawndale. Lawndale was the next step up from the Maxwell Street ghetto for many Jewish immigrants as they ascended into the middle class.

The family resided only two blocks from Douglas Park, a beautifully landscaped urban refuge with huge flowerbeds, lagoons, boat rentals, and free summer band concerts. Adrian attended nearby Theodore Herzl Public Grammar School whose student body was predominantly Jewish.

He graduated from Herzl in February 1919 and went on to attend John Marshall Public High School along with his grade school classmates. Filled with aspiring immigrant sons and daughters, Marshall High was extremely competitive. Some of the friendships he developed there endured throughout his lifetime.

Tragedy struck early in Adrian's life. A month after his sixteenth birthday, Minnie came down with nephritis. The young woman was engaged to be married. The infection took her life at the tender age of twenty.

Elias and Fannie weathered the tragedy together. Both had survived the loss of their mates. Now, they grieved as one.

Adrian graduated from Marshall High at the age of sixteen. As his high school years came to a close, America's doors of opportunity offered an opening for the struggling immigrants' son. He took home the John Marshall War Memorial Alumni Scholarship from his graduation exercises. Elias and Fannie felt a rush of pride. The scholarship would launch their son's studies at the University of Chicago.

Soon after the graduation ceremony, the family relocated again. Their new neighborhood was Chicago's Humboldt Park, a mostly middle class immigrant area inhabited by a mix of Poles, Germans, Italians, Ukrainians, Russians, and Scandinavians. Jews comprised about a quarter of the community. The Alberts took up residence a few steps from Division Street, the nucleus of the Jewish community and the area's principal commercial artery.[15] They moved into an apartment above their storefront fish market on a busy thoroughfare, California Avenue.

They lived within sneezing distance of another beautiful public park. The 207-acre Humboldt Park featured boating, bicycling, and swimming. Theodore Herzl Hall, a community center where meetings, classes and dances were held, sat in the middle of the park.

To the would-be college student, his new address came with an added benefit—Humboldt Park's elevated train station. It provided a fast track away from the drudgery of running a fish market toward his future at the University of Chicago.

Not that he intended to leave his family behind. His was a close-knit family. Over the years, he maintained close ties with his brother, who survived childhood despite bearing the rather feminine name "Joy," and his half-sister Esther.

In the 1930s Esther and her husband bought a summer cottage at the edge of a lake, where they had their own dock and a rowboat.[16] On visits with Esther and her family, Adrian grew to love the great outdoors and took up fishing as a hobby. There was no better way to rest his mind and restore his energy than sitting in a boat, waiting for a tug on his line.

Perhaps fishing was what kept him so down-to-earth. After marrying, he would tell his wife, "I don't want to vacation at any hotel too fancy for me to carry my fish through the lobby."

He approached fishing with the same tenacity he applied to the study of mathematics. Almost invariably, he landed his catch.

Adrian's future bride had her roots in another part of Europe. During the 19th century, the Davidovics family resided in the central European town of Szandrova. The town lay nestled among the foothills of the Carpathian Mountains. At the time, the area belonged to the Austro-Hungarian Empire under the rule of Emperor Franz Josef.

The sparsely populated area had no public school. A government-paid teacher came to town one day a week, but Jews were not permitted to attend classes. Mordecai Davidovics, the town rabbi, instructed the Jewish boys in reading of the Talmud.

Israel Davidovics was the third of five children born to Mordecai and his wife, Yitta. At the age of twenty, Israel set sail for America from Bremen, Germany aboard the "S. S. Friedriech d Grosse." Upon reaching port at Ellis Island in 1901, Israel Davidovics became Samuel Davis.[17]

Rivka Fuchs lived in a village adjoining Szandrova, where her father, Wolf Fuchs, earned his living in the lumber business. Rivka's brothers, like other Jewish boys in the area, studied under the tutelage of Rabbi Davidovics. Rivka's mother, Frieda, died when she was quite young, and Wolf married a widow with children some years older than Rivka. Their combined families boasted eight children in all.

Her brothers were large, strapping boys. That was a good thing when anti-Semitic boys from the surrounding area came calling with bricks and bats. In 1902, her half-sister married, and the couple decided to immigrate to America. Rivka seized the opportunity to leave the area. She was sixteen when she arrived at Ellis Island and anglicized her name to Rebecca Fox.

Despite the ocean that separated them, Samuel dared not choose a wife without consulting his father. No matter where Samuel resided, his prospective wife would have to withstand his father's scrutiny regarding her morals and family background.

Samuel's initial choice was soundly rejected, as the young woman's reputation had already traversed the ocean. However, Mordecai welcomed Samuel's next selection, Rebecca. One can only imagine that she, in turn, was proud to join the family of the learned gentleman who had instructed her brothers. The couple wed in 1904.

It was a good choice. Rebecca proved to be a sweet-tempered, capable person. She bore Samuel six children, all of whom revered this family matriarch. Her children's friends often sought her out in lieu of their own parents for advice on personal matters. The warm, caring woman was known for spearheading efforts to help those in her community who had fallen upon hard times.

Samuel and Rebecca resided in Buffalo, New York when their firstborn arrived in 1906. They named her Frieda, after Rebecca's mother.

When Frieda was three years old, the Davises moved to Chicago. Samuel eked out a meager living for his family as a tailor. As a young girl, she owned only two dresses—one for the winter months and one for the summer. She dreamed that, some day, she might have two dresses for each season.[18]

Frieda (her friends called her "Fritz" or "Curly Fritz" on account of her extremely curly hair) soon learned to sew, and grew to enjoy the pastime. She balked at the rule in her observant Jewish household forbidding her to sew on the Sabbath.

When America entered World War I in 1917, Samuel Davis' finances improved. He and a partner owned a thriving business making uniforms for American servicemen.

There was strong support for the war throughout the country. Frieda and her classmates sewed homemade gifts to send the soldiers. Among other items, they made fingerless gloves to keep the soldiers' hands warm without hampering their ability to handle weapons.

Radios first became popular in American households during World War I. One day, she came home from school to find her father and oldest brother in the backyard, winding wire around an empty cylindrical oatmeal box. The Davis family was handcrafting their first crystal radio set. It surprised her when she saw that it actually worked, although it was difficult to know which station was playing on the jerry-rigged contraption.[19]

She attended Tuley High, located virtually across the street from her family's rental apartment in the Humboldt Park neighborhood. Her friends threw a Sweet Sixteen party for the popular student. In January 1922, as graduation day approached, her fellow Tuleyites scribbled a variety of tongue-in-cheek advice in her autograph book.

A series of verses reflected the views of young women during "the roaring twenties" on their role in society. Most of them had marriage on their minds. A close friend left these words of wisdom: "Whatever you do,/Don't stay an old maid."[20] Another made this optimistic prediction: "Davis is your name/Single is your station/Happy is the man/Who makes the alteration."[21]

A few of the verses reflected a growing sense of liberation among women of that era. Her cousin, also a classmate, added a pessimistic note: "Take my advice/And don't fall in love."[22] Still another advised more drastic action: "When you are married/And your husband gets cross,/Pick up the broom and say/'I'm the boss.'"[23]

Frieda was in no rush to marry. A quick learner who loved books, she gravitated to the field of education.

After high school, she went to work, intending to save enough money to attend teacher's college. In the meantime, she remained at home, living with her parents and five siblings. It was a crowded but happy household, and there was no pressure to leave the nest.

She found a job as a bookkeeper and became adept at adding columns of figures. The efficiency-minded bookkeeper could add two columns at a time faster than the adding machines at her office.[24]

Chapter 2

Birth of a University in a New City

The city that gave birth to a great university grew out of a stretch of low, soggy flatland at the edge of a receded glacial lake. During their explorations in 1673, Jesuit missionary Jacques Marquette and fur trader Louis Joliet saw the potential of a swampy site at the foot of Lake Michigan. There, the waterway of the Great Lakes reached farthest into the virgin territory of the Great West. Although the aromatic onion bogs appeared unfit for human habitation, Joliet believed that if a canal were built connecting Lake Michigan to the Mississippi River, this location would control the North American continent.[1]

Joliet informed the Indians they met on their journey that God had sent his companion to teach them the "true faith."[2] The following year, Marquette returned to the site, where he founded a mission. But Marquette's health was failing and he died six months later.

Indian tribes held sway over the territory. A handsome entrepreneur named Jean Baptiste Point du Sable befriended the natives, took an Indian maiden as his wife, and made the area his home in the latter part of the eighteenth century. Du Sable, son of a French father and an African slave mother, was born in Haiti. The well-educated Haitian, a jack of many trades, built the first permanent non-native settlement in the territory.

The federal government wanted to control the strategic location. In the summer of 1795, General Anthony Wayne defeated twelve Indian tribes in a fierce battle. Afterwards, both victor and vanquished entered into the Treaty of Greenville. The Indians ceded a six square mile tract of land at the southwestern shore of Lake Michigan to the federal government. The government incorporated the tract as a town in 1833, giving it the Indian name "Chicago," meaning "wild onion."

Within a few years, the town had mushroomed to more than double its original size. The growing population formed a committee to apply for a city charter. The state legislature granted a charter to the City of Chicago in 1837.

Joliet proved to be correct about the value of the area. In just a few decades, Chicago became a thriving industrial center for grain, lumber, and meatpacking. The city's location as a transportation hub by water, and later, by railroad, fueled the growth of its economy. Huge grain elevators dominated the skyline. European immigrants flocked to the area to find work. Chicago epitomized the flowering industrial age in the nation's new Midwest. Entrepreneurs like William Butler Ogden, Cyrus McCormick, Philip Armour, Potter Palmer, and Marshall Field invested their capital and hard-driving work ethic into building the metropolis.

Contrary to the view of some historians, the city never wanted for intellectual and artistic culture. At the time of its incorporation, Chicago boasted a bookstore, theater, and newspaper, and offered regular concerts.[3] Within a few years, Walter Newberry established the Young Men's Association, which served as a privately run library.

Less than two decades after Chicago became a city, plans were in the works to build a university. The University of Chicago owes its origins to the United States Senator from Illinois, Stephen A. Douglas. In 1856, Douglas offered to donate ten acres of land for a university in the city.[4] Rev. John C. Burroughs, a Baptist pastor, seized the opportunity to secure the gift for his denomination after a Presbyterian group failed to act on the offer.

Within a year's time, the University of Chicago received a charter by act of the Illinois legislature. The first order of business was establishing a subscription fund to erect a building.

As luck would have it, subscriptions for the building fund were thwarted by the stock market panic later that year. The Great Chicago Fire of 1871, the panic of 1873, and the fire of 1874 further hindered the University's ability to meet its fundraising goals.[5] To finance construction, the Trustees were forced to encumber the University's land and buildings with a heavy mortgage debt.

One other factor impeded the University's ability to raise funds. Two years before agreeing to donate the land, Douglas had placed himself on the wrong side of history by introducing legislation that effectively nullified the Missouri Compromise. The Missouri Compromise, which excluded slavery from all territories north of the Mason-Dixon line, was sacred to northerners. Douglas' popularity in the North plummeted and remained at its nadir for a period of several years until he allied himself with forces seeking to exclude slavery from new territories.[6]

Because Douglas was so closely identified with the University, the public's hostility toward him made fundraising doubly difficult. In 1857, detractors charged that Douglas had contributed the land in order to enhance the value of his surrounding property. Fearing that his involvement would jeopardize the future of the institution, Douglas wrote to the Trustees, offering to take back the property and contribute cash instead.[7] The Trustees declined his offer on the ground that the donated land was the best site in the city.

Unlike some of the nation's top schools, Chicago did not long remain a "boys club." The forward-looking institution opened its doors to women students in 1873. The quality of education at the struggling school was high and many of the city's most distinguished citizens sent their sons and daughters to classes there. Despite its financial setbacks, the first University of Chicago enrolled approximately 5,000 students in its preparatory, college, and law departments during its brief lifespan.[8]

The University limped along financially for close to three decades. By the spring of 1886, the mortgage on university property had been foreclosed. The Trustees strove unsuccessfully to raise enough funds to redeem it.

Chicago clergymen turned to a faculty member of the Baptist Union Theological Seminary in the Chicago suburb of Morgan Park to lead them out of the morass. They began referring to William Rainey Harper as the "providential man" to head the University.[9] In April, the University's Board of Trustees elected him President of the University.

Harper came from an unlikely background for a role as head of a Baptist-led institution. Born of Scotch-Irish parents, he began life as a Presbyterian. He joined the Baptist denomination in 1876 only after accepting a position on the faculty of a Baptist institution, Denison University. Harper became an expert on the Old Testament after moving to the faculty of the Theological Seminary.

Seeing no hope for the success of the University of Chicago, he declined the post as President. Instead, he accepted a chaired professorship in Hebrew at Yale University. The first University of Chicago closed its doors permanently two months later.[10]

Chicago suffered a crushing blow due to the Great Chicago Fire. The fire consumed 2,200 acres and 18,000 buildings, devastating the city and its economy. About a third of the city's 300,000 residents were rendered homeless.

The city began to pull itself up by its bootstraps, determined to rebuild. Less than two years after the fire, Chicago constructed an Inter-State Exposition Building and began hosting fairs.[11]

Chicago's first Art Institute opened its doors in 1885. The work of French impressionists was displayed in a glorious Romanesque building designed by architects Daniel Burnham and John Wellborn Root.[12] The institution catered to the city's working class as well as its moneyed entrepreneurs by offering free admission on Sundays.[13]

Only four years later, Adler and Sullivan's masterpiece was completed. The Auditorium Theatre Building, with its magnificent ornamental arches, gilded plaster relief, and nearly perfect acoustics, opened to thrilled crowds from all economic strata.[14]

The Chicago of the mid-nineteenth century was tiny compared to the city it would eventually become after annexing many of the adjoining suburbs. In 1853, lawyer-entrepreneur Paul Cornell purchased 300 acres of lakefront land south of the city. He named his purchase Hyde Park, after London's manor of Hyde.[15]

Cornell billed Hyde Park as a resort town away from the congestion and strife of the big city. Hyde Park enacted ordinances forbidding social disturbances and vice. Fines of up to $100 were imposed for noisy disturbances of the peace.[16] He built an elegant hotel, and Hyde Park became a tranquil retreat for Chicago's business and professional men and their families. Abraham Lincoln's wife and children retreated to the Hyde Park House hotel after the president's assassination.[17]

Cornell and his fellow investors successfully lobbied the Illinois legislature in Springfield for a Park Bill. In 1868–69, the Park Commission used bond revenues to purchase Jackson Park along the city's lakefront, Washington Park farther west, and a connecting strip of parkland stretched "midway" between the two.

Cornell's original town of Hyde Park later merged with some twenty-three other communities to form a 48 square mile township.

The township was incorporated as the Village of Hyde Park in 1872 pursuant to an Act of the state legislature.

Despite the protestations of Cornell and many of his followers, Chicago swallowed up the tranquil Village of Hyde Park by annexing it in 1889. The neighborhood known as Hyde Park today comprises only about three square miles of the former village.[18]

After the first University of Chicago closed its doors, Baptists in Chicago felt humiliated over the loss of "their University." However, leaders of Chicago's Baptist community refused to accept defeat. One of the most determined of these was Thomas Goodspeed, who took it upon himself to seek reincarnation for the institution.

Goodspeed and his fellow church leaders managed to engage the interest of John D. Rockefeller, a powerful figure who owned a flourishing oil business. While some viewed him as a ruthless oil baron, Rockefeller, a devout Baptist, was amenable to the idea of establishing a new university under Baptist control.

While he leaned in favor of locating the university in Chicago, the philanthropic easterner was unwilling to take the next step without input from the American Baptist community at large. It took the formation of a national organization, the American Baptist Education Society, to survey the educational needs of American Baptists and come up with a plan of action. The fact that Chicago had trebled in size within two decades and had become the nation's second largest city weighed heavily in the Society's decision to favor Chicago as the site of a Baptist-controlled university. A spokesman for the Society argued that the university should be located in Chicago because the city was "the social, financial, literary, and religious eminence in the West....the fountain of western life."[19]

The Society joined forces with the Chicago group in appealing to Rockefeller to support the project. Rockefeller initially pledged $600,000 for a university in Chicago on condition that the Society would raise the balance of a one million dollar initial subscription.

In the course of seeking matching money for the subscription, fundraisers reached out to non-Baptists and eventually, to non-Christians as well. Chicago's Jewry made substantial contributions to the school.

The new University of Chicago received its charter in 1890. Its policies were enlightened ones for that era. The charter called for Baptist control of the institution—two-thirds of the University's trustees were required to be members of Baptist churches. However, given the diversity of the university's donors, the charter forbad the use of any religious requisite for selection of professors or admission of students.

In addition, the new institution, like its predecessor, opened its arms to women students. The new charter mandated that opportunities be provided to students of "both sexes on equal terms."[20]

Location was another important criterion. The charter stressed that the university must be located within the confines of Chicago and not in a suburban village.

Goodspeed and his fellow church leaders had hoped to rebuild on the historic site of the old university, but discovered the cost of the original site had skyrocketed. They eventually found an ideal location in the newly annexed Hyde Park area—a plot of unoccupied ground fronting on the Midway Plaisance between Washington and Jackson Parks. Chicago's leading merchant, Marshall Field, owned the ten-acre site. It was part of an 80-acre tract that Field had purchased a decade earlier for about $79,000.[21]

The fundraisers paid a friendly call on the merchant to discuss the advantages of donating the land for their project. They approached the task with much trepidation, as Field had a reputation for turning a cold shoulder to charitable appeals. Reportedly, when one of Field's closest friends had suggested that he ought to found a great university in Chicago as a monument to the city where he prospered so phenomenally, Field replied, "other men might build monuments if they wished.... [It's] very easy to give away other people's money."[22]

The shrewd businessman received the Baptist group cordially, but demurred that he would need six weeks to look into his company's finances. At the end of the six-week hiatus, they pressed him for a decision, arguing that his commitment would help them immensely in persuading others to contribute. Seeing the potential impact of his promise, Field agreed to pledge one ten-acre parcel on condition that it not be counted as part of the matching money needed for Rockefeller's subscription. A week later, he sold the Society an option on an adjoining ten-acre parcel. Field set the cost of the second parcel at $132,500, nearly double what he originally paid for the entire 80-acre tract.

Goodspeed and Rockefeller had actually been in close communication since the early 1880s. Before becoming involved with the University of

Chicago, Goodspeed had assumed the role of chief fundraiser for the Baptist Union Theological Seminary. Rockefeller was one of the Seminary's chief benefactors. As a supporter of the Seminary, he came to know the scholarly professor who held a chair in Hebrew studies there.

It was in Rockefeller's correspondence with Goodspeed about the Seminary that the subject of a new university in Chicago first arose. Afterwards, the pair began an almost conspiratorial effort to court Harper as head of the university.

When he learned of Yale's interest in Harper, Rockefeller personally wrote to Goodspeed, alerting him to the possibility that Harper might leave the Chicago area. As previously mentioned, Harper did accept Yale's offer in 1886. But Harper's apparent unavailability as a candidate made him all the more appealing to the wealthy oil baron who was accustomed to having his own way in business matters.

In October 1888, Harper wrote Goodspeed to relate a curious incident. "I spent last Sunday at Vassar College….Much to my surprise Mr. Rockefeller was there….What his purpose in going to Vassar was is not quite certain. He seemed to have nothing to do there except to talk with me."[23]

Harper was torn between his love for biblical studies and the good he could do by assuming the post. He had no desire to leave Yale to become involved in a second-rate institution. From the beginning of his involvement, he did his utmost to ensure that the goal would to establish a great university, not a mere college.

Rockefeller was unwilling to commit to the project without Harper. Rockefeller saw a dynamic leader in the Yale professor despite the fact that Harper's outward appearance and demeanor in no way resembled his own. The professor was "pudgy in physique, scholarly in tastes, ebullient in temperament, and conspicuously outgoing in manner."[24] Rockefeller finally added a million dollars to his initial pledge to persuade Harper to leave Yale. The contribution allayed some of Harper's concerns that the school lacked the resources to meet his expectations, and he accepted.

Rockefeller, who contributed the majority of the university's endowment, was widely viewed as its "founder." His endowment enabled the new university to survive a second major stock market crash. Even the Depression of 1893 could not stifle the University's progress.

Despite his role as "founder," he declined a seat on the Board of Trustees. He placed his trust in Harper to run the institution of higher learning.

While formed by Baptists, the university had ties to other religious bodies as well. The American Baptist Education Society appointed members of Baptist, Unitarian, Universalist, and Jewish congregations to sit on the Board. The 21-member board included only a single clergyman. The rest were judges, lawyers, industrialists, wealthy men, and one academic—Harper.

The fundraisers had promised donors that the new Board would substitute businessmen for ministers and do everything possible to avoid the mistakes of the past. Failure was not an option.

The 1890s ushered in a second development unfolding side by side with construction of the University. The federal government anticipated the four hundredth anniversary of Columbus' voyage to the New World by planning a world's fair to commemorate the occasion.[25] Congress passed an Act creating the World's Columbian Exposition less than a year after Hyde Park became part of Chicago.[26] Chicago, still struggling to recover from the fire, saw the world's fair as an important opportunity, and competed vigorously for the right to host it.

Chicago won the bidding, much to the chagrin of competing New Yorkers who considered Chicago a cultural and intellectual backwater. Visitors would throng to the grand fair in the city of the future, being built anew from the ground up. The fair would reflect, not the city sullied by the blackened fumes of industry and catastrophe, but the hopes of a nation seeking an idyllic future. All of the main fair buildings would be sparkling white.

Adjacent to the fair, the process of transforming a former fledgling Chicago institution into a great new university was taking place. The Fair of 1893 extended along the southern flank of the University. While a fanciful world of shimmering white Beaux-Arts buildings, lagoons, and Venetian canals emerged along the lakefront and grew westward down the Midway Plaisance, majestic Gothic Revival structures rose Phoenix-like from the ashes of the old college.*

---

*   The first University of Chicago was actually located farther north, at 34th Street and Cottage Grove Avenue.

Chicago architects Burnham and Root were picked to direct the fair project, and world-renowned landscape designer Frederick Law Olmsted was designated as landscape architect. Burnham and Olmsted selected Jackson Park and the adjoining Midway as the future fairground because of the park's lakefront location. They reasoned that any exposition honoring Christopher Columbus should incorporate water in its design.

During 1891–92, Olmsted and his crew undertook a massive makeover of Jackson Park's marshy flats mired in reeds and scrub vegetation.[27] Root lived on the site during construction, as did the sculptor, Augustus Saint-Gaudens.[28]

The two developments shared some of the same planners. Henry Ives Cobb, author of the master plan for the new university campus, designed several of the Columbian Exposition's most beautiful buildings. Cobb presented a unique aquatic display in his Fisheries Building, a confection of ornate circular pavilions. The French Gothic Marine Cafe with its many spires and witches' caps and the Indiana Building with its red towers were also his creations. Cobb's recreation of "A Street in Cairo" along 59th Street was one of the Fair's most popular attractions.

Charles T. Yerkes, a transportation magnate who believed the end justified the means when it came to monopolizing Chicago's transportation system, served on the board of the corporation formed to finance the fair. In Yerkes' view, bigger was better. To burnish his tarnished image, he agreed in 1892 to donate funds to the University for an astronomical observatory featuring the world's largest telescope.[29]

The Fair, which opened in the spring of 1893, enhanced Chicago's reputation as a cultural center. It brought great architects and cultural icons to the city. At a series of symposia associated with the Exposition, audiences thrilled to speeches by John Dewey, Julia Ward Howe, Samuel Gompers, Clarence Darrow, historian Frederick Jackson Turner, and other greats.

In the fall, the extravaganza was over. Afterwards, in a burst of literary excess, a well-known journalist of the day declared that the Fair had been "the greatest event in the history of the country since the Civil War."[30] According to one account, the Fair had "gathered up and presented to millions of people much that was new and challenging in every field of human activity." Guidebooks described it as "the marvel of the century, a comprehensive picture of civilization, the culmination of all the progress made by the nations of the earth in the centuries that had passed, and the gathering together of all that science, art, and ingenuity had produced for the benefit of mankind."[31]

While the World's Fair brought a great deal of attention to Chicago as the nation's cultural "second city," its magnificent architecture was constructed as a "luxuriant midsummer efflorescence, born to bloom for an hour and perish."[32] Most of the Fair's structures, covered with a fragile mix of plaster, cement, and jute, were demolished after the event.

The Gothic works of art rising on the adjoining campus, on the other hand, were built to last. Their medieval designs, rooted in Gothic geometry and proportion, would be constructed of solid stone.

Chapter 3

The First Two Generations at the Mathematics Department

Classes began at the new university in 1892. President Harper passed over more experienced academics and recruited the 29-year old Eliakim Hastings Moore to become acting head of the Department of Mathematics. Moore had earned his doctorate at Harper's alma mater, Yale University, a decade after Harper received his.

Moore needed convincing, since he was content to remain at a well-established school, Northwestern University, where he had taught for several years. Harper managed to entice the youthful scholar into accepting by elevating him to full professor and promising a low teaching load that would allow him to concentrate on research.

Harper's confidence in the young man's organizational ability soon bore fruit. Moore was instrumental in organizing a Mathematics Congress held in association with the World's Fair.[1] As a result, Chicago's brand new Mathematics Department played host to mathematicians from across the United States and several foreign countries, thereby imparting a certain legitimacy to the fledgling Department.[2]

A number of mathematical papers were read at the Congress, including a paper by a young German mathematician named David Hilbert containing a novel proof that had begun to shake up the mathematical establishment in Europe. Hilbert's paper on "invariant theory," presented by his German colleague, Felix Klein, created a sensation at the international gathering in Chicago. Hilbert had managed to prove that there was a finite set of building blocks at the basis of mathematical equations.[3]

The "Chicago Mathematics Congress" was so successful that it inspired the mathematicians who had shared the experience to begin planning the first of a series of quadrennial international congresses of mathematicians.

Klein stayed over in the Chicago area after the Congress for another event—he was scheduled to deliver a set of lectures at a Mathematics Colloquium.[4] His lectures were to be held for one week at Northwestern University's Evanston campus and a second week at the University of Chicago. However, the remnants of a harsh midwestern winter interfered with the plan. As the heavy ice and snow build-up melted under the onslaught of spring rains, a 2-foot flood invaded Chicago's campus and the colloquium had to remain at Northwestern.

Klein was a controversial figure. On the one hand, he earnestly and eloquently promoted positive reforms in mathematics education. On the other hand, he espoused notions about the superiority of the Teutonic race that would later find their way into Nazi doctrine. During his remarks at the Colloquium, he speculated that members of the Teutonic race possessed a mathematical intuition superior to that of "the Latin and Hebrew races."[5] Despite the somewhat flawed content of his lectures, the Colloquium's success led to an American tradition—the annual lecture series delivered by a top mathematician designated as "Colloquium Lecturer."

Moore rounded out the new Department with two outstanding German mathematicians: Heinrich Maschke, known for his work in group theory and differential geometry, and Oskar Bolza, who specialized in the calculus of variations and other topics in mathematical analysis. Bolza's text, *Lectures on the Calculus of Variations*, became a classic in the field.[6]

Under Moore's guidance, Chicago grew into a powerhouse of mathematical prowess. The three original members of Chicago's mathematics faculty soon attracted the notice of scholars for the quality of their research, and Chicago's Mathematics Department became recognized as the number one department in the country.

In 1896, the Department awarded its first Ph.D. in mathematics to a brilliant young Texan, Leonard Eugene Dickson, who completed his dissertation under Moore's supervision.[7] Only four years later, Moore added his former doctoral student to Chicago's mathematics faculty.

Adrian Albert belonged to the second generation of mathematics students at the University of Chicago. The sixteen-year old boy with a large crop of curly hair entered the University as a freshman in the fall of 1922, armed with nothing more than a powerful will to learn and a first year scholarship awarded by his high school.

As a youngster, he had developed a fondness for gadgets. From an early age, he built radio sets as a hobby. He learned about the most advanced radio circuits and incorporated them into his radios. The hobby helped support him during college. To earn spending money, he worked as a radio repairman on Saturdays and during the summers.

The first-year college student did not mind the commute across town from his family's latest residence in Humboldt Park. Among the members of his immediate family, he alone had the opportunity to attend college. He was doubly grateful for the chance to continue his studies when he learned that the star pupil in his graduating class at Marshall High, a young woman, dropped out of the University of Chicago to help support her family.

Although he lived amid the hustle and bustle of his family, it did not impede his studies. When Adrian studied mathematics, he entered his own domain. His concentration became so intense that he neither saw nor heard those around him.

The bulk of support for his education came from a series of scholarships—all based on merit. At the end of his first year, he received a University of Chicago Honors Scholarship awarded to the top 20 students in the entering class. He earned scholarships for the last two years of college by winning competitive examinations given by the Mathematics Department.

Mathematics majors at the University in 1922 were required to complete at least nine one-quarter courses in the subject. He was not content with such a limited menu of mathematics. The undergraduate loaded up his schedule with twenty courses in mathematics, nine of which were graduate level classes.

L. E. Dickson's classes in Theory of Equations and Number Theory opened up a whole new world to the budding mathematician. Dickson was fond of quoting the great Karl Gauss for the proposition that mathematics is the Queen of the Sciences and the Theory of Numbers is the Queen of Mathematics. Dickson's class in Number Theory introduced Adrian to the notion that fascinating relationships exist between the ordinary whole numbers.

A course on functions taught by Gilbert Bliss expanded the college student's horizons by introducing him to the concept of mathematical mappings.

Another teacher, Herbert Ellsworth Slaught, one of the principal founders of the Mathematical Association of America, demonstrated the value of making a contribution to mathematics apart from conducting pure research. The significance of Slaught's leadership role in improving mathematics education was not lost on the undergraduate.

Adrian studied differential and integral calculus under the outstanding mathematician, Ernest Wilczynski, who earned his doctorate in Berlin. Moore invited Wilczynski to join the faculty after Oskar Bolza returned to Germany in 1910.

Wilczynski, a prolific research mathematician, began his career as a mathematical astronomer. He taught a basket of courses at Chicago, including projective differential geometry, differential equations, solar rotation, spiral nebulae, and ruled surfaces.[8]

It was apparent to those around him at Chicago that Adrian broke the mold. One of his college classmates later recalled:

> About all I remember is that Ade was too brilliant for the rest of us...I never really got close enough to Ade to observe any particular eccentricities or idiosyncrasies in him. I suspect that he was a particularly sensitive person and that sensitivity laid him open to being hurt rather easily. So this kept him from opening up in many respects and protected him from becoming the subject of anecdotes. My memory of him then was as a very gentle, kind, and even sweet person as well as terrific intellectually.... I still can feel warmed by the vivid memories of the twinkle in his eyes, that indescribably sweet smile, and that soft, kindly chuckle.
>
> The other strong recollection I have is of the burning intensity of his love of mathematics, his restless intellect, and his determined pursuit of research.[9]

In the spring of 1926, he graduated Phi Beta Kappa from the University of Chicago, earning a Bachelor of Science degree in mathematics with a minor in physics. The majority of students in his graduating class were men, but the class contained a large contingent of women. All races were included in the Class of '26.

He knew early on that he was headed for a career in academe. Given the necessity of selecting an institution for his graduate studies, he considered his options. The gifted student had the potential to secure an academic scholarship anywhere in the United States.

A 1925 report prepared for the American Association of Colleges ranked Chicago's Mathematics Department first among mathematics departments in American graduate schools.[10] Chairman Moore had managed to assemble a top-notch faculty.

Moore himself was an inspiring mathematician. Perhaps his greatest contribution was introducing a new methodology into American mathematics—treating mathematical concepts as abstract "objects" not bound to any particular number system. This abstract approach influenced his student, Dickson, and many of the mathematical progeny spawned by Moore and Dickson.[11] Moore, who published 74 mathematical papers between 1885 and 1926,[12] toiled in all three branches of mathematics: analysis, geometry, and algebra. At Chicago, he taught classes in "general analysis" (Moore's unique brand of function theory), finite fields, simple groups, algebraic geometry, number theory, and integral equations.[13]

He received recognition for several notable advances. In a paper delivered at the 1893 "Chicago Mathematics Congress," Moore discovered a new class of finite groups.[14] Later, he developed an important theorem relating to finite fields.[15] In 1902, Moore's paper "On the Projective Axioms of Geometry" showed that the axiom system developed by Hilbert contained redundant axioms.[16]

Dickson, a stellar figure in linear algebras and number theory, also made major contributions to the theory of finite fields, linear groups, and algebraic invariants. After Moore lured him to join Chicago's mathematics faculty in 1900, Dickson remained on board until 1939.

Dickson produced an impressive volume of research—267 mathematical papers and 18 books. His three volume *History of the Theory of Numbers* could have been a life's work in itself. One of his best-known works was the book, *Algebren und ihre Zahlentheorie*, which laid the groundwork for future research on linear associative algebras.[17] Dickson's work on associative algebras earned him a well-deserved reputation as one of the founders of the American school of algebra.

Bliss, another former first generation University of Chicago mathematics student, earned his doctorate under Oskar Bolza, whose inspiring lectures on the calculus of variations influenced his career.

During World War I, Bliss worked on ballistics and designed firing tables for artillery. At a U.S. military testing facility in 1918 he effectively applied his expertise in the calculus of variations to solving problems involving the trajectories of missiles affected by wind, changes in wind density, and rotations of the earth.[18]

Bliss occupied a prestigious place on the American mathematical scene. Among other honors, he received the Chauvenet Prize from the Mathematical Association of America for his article on *Algebraic Functions and Their Divisors*. In 1921, he became President of the American Mathematical Society. Besides more traditional mathematics courses such as Calculus of Variations and Differential Equations, he taught a class called "Exterior Ballistics" at Chicago.[19]

Moore wooed Bliss away from a teaching position at Princeton University in 1908 after the talented teacher, Heinrich Maschke, died at the age of 55. Bliss would succeed Moore as Chair of Chicago's Mathematics Department in 1927 and keep the post until the eve of America's entry into World War II.

Given the stellar array of scholars in Chicago's mathematical constellation, Adrian had no trouble making up his mind about where to attend graduate school. There was no reason to look elsewhere when the best was available right at home.

He managed to afford graduate school at Chicago with the aid of an Honors Scholarship, followed by a series of research fellowships. He taught trigonometry during his last year of graduate school as required by his fellowship.

In graduate school, he was permitted to study under the august E. H. Moore for the first time. Moore's approach to graduate-level teaching differed radically from the standard lecture format of most mathematics professors. "Moore's graduate teaching was done in a research laboratory setting. That is, students read and presented papers from journals, usually German, and tried to develop new theorems based on them. The general subject of these seminars was a pre-Banach form of geometric analysis that Moore called 'general analysis.' "[20] He took a sequence of five courses from Moore, all listed as "General Analysis."

Moore had a profound influence on the 20-year old graduate student. Adrian later credited the charismatic teacher with arousing his interest in pursuing abstract mathematics.[21]

During his graduate studies, he added another 24 mathematics classes to his plate. Among them was Bliss' course on "Calculus of

Variations," the skill set that aided Bliss in his war work on ballistics. He also found time for a class in "Modular Systems" taught by E. T. Bell, who visited Chicago for the summer of '27.*

Adrian toiled on both his master's and doctoral dissertations under Dickson's supervision. While Moore's analytical approach drew him to the field of abstract mathematics, he saw his future in Dickson's brand of algebra. Dickson's 1923 tome, *Algebras and Their Arithmetics*, and his 1926 paper, *New Division Algebras*, featured cutting edge mathematics that stirred his creative energies.[22]

Dickson was as colorful as he was brilliant. Looking back at the 1920s, one of his students gave this assessment of Dickson, the teacher:

> In the conventional sense Dickson was not much of a teacher. I think his students learned from him by emulating him as a research mathematician more than being taught by him. Moreover, he took them to the frontier of research, for the subject matter of his courses was usually new mathematics in the making. As...[another classmate]† said, "He made you want to be with him intellectually. When you are young, reaching for the stars, that is what it is all about." He was good to his students, kept his promises to them and backed them up. Yet he could be a terror. He would sometimes fly into a rage at the department bridge parties, which he appeared to take seriously. And he was relentless when he smelled blood in the oral examination of some hapless, cringing victim. He was an indefatigable worker and in public a great showman, with the flair of a rough and ready Texan.[23]

Another 1920s student recalled, "...good old Leonard would dump his pipe ashes into the wastebasket and get a fire started in the waste paper. Whereupon he would delicately place the waste basket out in the hall until it went out!"[24]

In light of his family's strained finances, Adrian was anxious to complete his education and begin earning a living as soon as possible.

---

* Bell later became well known for his lively biographies of famous mathematicians and books on the history of mathematics.

† The classmate, Antoinette Killen Huston, later became Adrian Albert's first doctoral student.

He wrote his master's thesis on the "determination" of linear algebras in two, three, and four dimensions.[25] Before the end of his first quarter as a graduate student, he had completed the thesis, and Dickson had accepted it.[26]

Adrian was well along in his doctoral thesis before formally receiving his master's degree in June 1927. He completed an acceptable doctoral thesis titled "Algebras and Their Radicals" by the fall of 1927, a mere fifteen months after graduating from college.

At that point, Dickson suggested an additional thesis topic—"determining" all normal division algebras in sixteen dimensions.[27] At first blush, Dickson's suggestion seemed odd in light of the fact that he had already approved Adrian's first doctoral thesis. The new assignment was a problem that had stymied mathematicians, including Dickson himself. Perhaps the doctoral candidate sensed that his teacher was passing the torch, and he readily agreed to tackle it.

Adrian dispatched the second problem within a few months. He stapled the two dissertations together, adding only the words "and Division Algebras" to the title page. His masterpiece, the passport to his future career, lay in those three inconspicuous words. According to one commentator, "In 1928, at age 22,...[A. A. Albert's] Ph.D. dissertation already stamped him as one of the outstanding algebraists of his day."[28]

He spent the summer taking the oral examinations for his doctorate while squeezing in three additional courses. He garnered a Ph.D. in Mathematics in August 1928 at the age of 22.

Adrian acknowledged Dickson's role in his career early and often. In his doctoral dissertation, he wrote, "To Professor Leonard Eugene Dickson who furnished the guidance, aid, and inspiration for the writing of this thesis, the author owes more than he can adequately express."[29] In his 1939 treatise, *Structure of Algebras*, he praised Dickson's expositions on structure theorems and described him as "the inspiration of all my research."[30] When Dickson died, it was Adrian who wrote his memoir.[31]

The algebraist Irving Kaplansky believed that Dickson's influence set the course for much of his pupil's subsequent research.[32] As a fellow schoolmate of Adrian's pointed out, "L.E. Dickson's students tended to identify themselves strongly as number theorists or algebraists. I felt this particularly in...A. [Adrian] Albert, Gordon Pall and Arnold Ross. All his life Albert strongly identified himself, first as an algebraist, later with mathematics as an institution and certainly with the University of Chicago."[33]

On the other hand, he was nobody's copycat. In a firsthand account of Adrian Albert as a student, W. L. Duren, Jr. wrote:

> I remember him as an advanced graduate student walking into Dickson's class in number theory that he was visiting, smiling and self-confident. He knew where he was going. Dickson was teaching from the galley sheets of his new *Introduction to the Theory of Numbers* (University of Chicago Press, 1929) with its novel emphasis on the representation of integers by quadratic forms. I think he requested Albert to sit in for his comments on this aspect. He was tremendously proud of Albert. I remember A cubed too with his beautiful young wife, Frieda, at the perennial department bridge parties. He had superb mental powers; he could read a page at a glance. One could see even then that as heir apparent to Dickson, he would do his own mathematics rather than a continuation of Dickson's, however much he admired Dickson.[34]

A.A. Albert looked to Dickson as his chief mentor at Chicago. Dickson not only taught the budding mathematician how to pursue cutting edge mathematics, he influenced and guided the direction of his disciple's career as a research mathematician.[35] Dickson had steered the young man on the right course by offering the dissertation topic that launched his career. Long after receiving his doctorate, he continued to confer with Dickson about his research.

While Dickson's mentoring influenced his mathematical research, Adrian looked elsewhere for a role model to follow as he progressed from novice to leader and change agent in the world of mathematics.[36] He looked to someone with a transcendent spirit, a kind and gentle disposition, and an ability to influence others for the betterment of his profession.

Chapter 4

The Early Years at Chicago and Princeton

Adrian Albert's mathematical career was nearly derailed shortly after entering graduate school. In February 1927, Elias died at the age of 72. It was a crushing blow to the shy, introspective 21-year old who felt he had lost his best friend. To make matters worse, the family began to question whether he should remain in school now that the mainstay of the family business was gone.

Elias was more than a good father to his firstborn son—he was Adrian's champion. Elias had done everything in his power to ensure that his son's future would unfold in front of a classroom, not behind a grocery counter.

Adrian coped with his grief by burying himself in his work. The tactic enabled him to persevere. It was a tactic he would need to draw upon again. Unwilling to become mired in a tragedy he could not undo, he focused his attention on the future.

He intensified his drive to complete his doctoral thesis and became rather withdrawn and solitary. Fannie urged him to get out of the house and socialize.

April rolled around before he finally acceded to his mother's wishes. He wandered over to a social at Theodore Herzl Hall, the local community center where eligible youth went to meet and greet each other. His eyes drifted across the room to a vision wearing a bright green dress. Taking in her peaches and cream complexion and finely chiseled features, he watched as the vision chatted with her friend. When the friend accepted an invitation to dance, the young woman sat alone. He seized the moment and asked her for a dance.

The couple danced together for the entire evening. He discovered that the vision's name was Frieda, and she was head bookkeeper for an automotive parts company. The company made hood ornaments designed to reflect a car-owner's personality. She learned that her partner's name was Adrian, and he aspired to become a mathematician. She was a good dancer; he was not. Frieda didn't seem to mind. "I made a choice about what was more important," she would say later.

At the end of the evening she went home and told her mother, "I met the man I'm going to marry." Rebecca looked anxiously at her daughter. "What does he do for a living?" Frieda was undaunted. "He's a student and he's going to be a college professor." To which Rebecca countered, "If you marry a professor, you'll starve!"

Although she valued Rebecca's counsel on most matters, Frieda would follow her own instincts. She felt certain she had made the right choice. Within weeks, the couple became officially engaged. Her friends were shocked. "You can't get married," her best friend protested. "He's unemployed and you don't even know how to boil a cup of water!"[1]

About eight months after they first met, the couple wed on a bitterly cold Sunday night, December 18, 1927. Theirs was a religious ceremony at Haddon Temple in their Humboldt Park neighborhood. Adrian looked dapper in his rented tux and Frieda was resplendent, the perfect flapper in her white satin gown with its scalloped hemline. After the wedding, guests celebrated at the bride's family residence— a third floor apartment over a storefront on west Division Street.

No one could fill the void in Adrian's life left by the loss of his father. The young man was not replacing a father; he was gaining a wife and life partner. Still, marrying so soon after losing his father was no coincidence—in matters of the heart, timing is everything. With Frieda at his side, he felt stronger than ever and ready to face any challenge.

Adrian had grown up in Jewish communities who revered the iconic figure, Theodore Herzl. Both his grade school and the community center where he met his bride bore Herzl's name. According to former Israeli Premier Golda Meir, Herzl was a reporter for a newspaper in Vienna who underwent a life-changing experience after being assigned to cover the trial of one Captain Alfred Dreyfus of the French general staff. In 1894, his superior officers at France's War ministry arrested Dreyfus for passing military information to Germany.

Meir explains that Herzl was "[s]hocked by the injustice done to this Jewish officer—and by the open anti-Semitism of the French army...[which led Herzl] to believe that there was only one possible permanent solution to the situation of the Jews."[2] Herzl later became known as the founder of the World Zionist Organization, and, as Meir put it, "the father of the State of Israel."*

Despite the fact that public buildings in his environs were named after a leading advocate of Zionism, Zionism was neither Adrian's cause nor his interest. Like a large segment of America's Jewry, he perceived an inherent tension between the American Zionist movement and the assimilation of Jews into the social and economic fabric of the nation. America was his homeland, and he felt an intense loyalty to it.

From the beginning of her husband's career, Frieda was there, cheering him on and helping to ease his way. She kept her bookkeeping job while he completed his doctorate. To conserve on expenses, the couple opted to live with his mother in the apartment above the Albert family's former fish market. The store itself had sat vacant and shuttered ever since Elias' death.

It was April when Adrian presented a paper to the American Mathematical Society based on the research contained in his doctoral dissertation.[3] Although he had yet to officially receive his Ph.D., the paper earned him a postdoctoral fellowship from the National Research Council.

Frieda joined his mother in the audience while he donned his cap and gown for the 1928 summer convocation. Rockefeller Chapel, where future convocations would be held, was two months away from completion.[4] Graduating students and their guests sat on uncomfortable folding chairs outdoors in Hutchinson Court for the convocation on that sultry August day.[5] But when Adrian marched up to receive his doctoral diploma, the discomfort he was feeling seemed a small price to pay.

Less than a month later, the couple sped off to spend his postdoctoral year at Princeton University. Next to Dickson, Princeton's J. H. M. Wedderburn was the foremost expert on associative algebra in

---

*   Dreyfus was finally exonerated in 1906.

the United States. Studying the methods of a master algebraist like Wedderburn was just what the young scholar needed to cap off his education.

The couple toured the historic town of Princeton and Princeton University's campus. The University's Nassau Hall, where the first Continental Congress met, had survived the Revolutionary War and two devastating fires.

The town of Princeton, once an overnight stagecoach stop between New York and Philadelphia, had grown into a well-to-do citadel of gentility. One might speculate it was in that rarified Eastern atmosphere that Frieda acquired her affinity for faculty dinners served on fine china rather than paper plates.

A few months before the Alberts got their first glimpse of Princeton's campus, a construction crew placed the finishing touches on Princeton University's cross-shaped Gothic Revival chapel.[6] When they left the Midwest, construction workers were close to completing Chicago's Rockefeller Chapel.[7]

Princeton University, like its Chicago cousin, evolved from a previous incarnation. Before 1896, Princeton was called the College of New Jersey. It became a university in 1896, four years after the new University of Chicago opened its doors. In 1900, Princeton awarded its first-ever doctorate degree in mathematics.[8]

When the Alberts arrived in 1928, ideas about constructing a mathematics building remained on the back burner at Princeton while, 700 miles away, Chairman Bliss sat poring over plans for Chicago's first mathematics building. In the absence of office space, Princeton's mathematics professors were obliged to pursue their research at home. Mathematics students and Research Fellows toiled in Princeton's neglected physics department building, Palmer Laboratory, where rats roamed freely among the library's stacks.[9]

During the year, Adrian drew inspiration from Wedderburn's elegant results and methods.[10] To all outward appearances, Joseph Maclagen Wedderburn and Abraham Adrian Albert were as different as any two human beings on earth. Born and raised in Scotland, Wedderburn never fully acclimated to life in America. He joined Princeton's faculty in 1909, but left to enlist in the British Army when World War I broke out, and did not return to Princeton until the end of hostilities.

Wedderburn remained a lifelong bachelor who spent his leisure time in the company of other bachelors. Students found him detached and aloof. At Princeton, he supervised no more than four doctoral students during his entire tenure. Adrian was one of the few scholars coming to work with the shy, introverted professor who received a great deal of support from him.[11]

Shared interests breached the chasm of disparities between the two men. Both had studied under L. E. Dickson in their formative years and excelled at Dickson-style algebra afterwards. Both loved the great outdoors. And both men devoured prodigious quantities of mystery stories.

In the ensuing years, Adrian corresponded with him, keeping Wedderburn abreast of advances in his research. Although Wedderburn, who suffered from severe depression, gradually withdrew from society, Adrian managed to visit with him on return trips to Princeton.

His year of exposure to Wedderburn served to further intensify the spark of interest in associative algebra that Dickson had ignited. That alone would have made his 1928–29 postdoctoral year worthwhile.

The year yielded yet another treasure for the postdoctoral fellow. It happened at the Princeton Mathematics Club, a weekly gathering of faculty, research fellows, and graduate students on campus. By chance, Princeton University's Solomon Lefschetz, a pioneering expert on algebraic geometry, was in the audience when Adrian lectured on his Ph.D. dissertation.[12] Lefschetz' mathematical intuition grasped something in the young man's exposition of associative algebras and their matrices that reminded him of a long unsolved problem arising from the work of a nineteenth century geometer, Bernhard Riemann.

After the lecture, the two men proceeded to discuss the problem for several hours in the course of wanderings through the streets of Princeton before either of them realized it was well past dinnertime. Adrian finally returned home to inform his anxious wife that he had just been initiated into a fascinating area of classical mathematics.†

That year, Lefschetz took over from Wedderburn as Editor of Princeton's journal, *Annals of Mathematics*. By raising the journal's standards, the dynamic Lefschetz was to play a major role in shaping it into one of the world's most prestigious mathematical publications.[13] During the fall of 1928, the *Annals* began accepting some of Adrian's research papers for publication.

---

† More about Adrian's work on the problem of Riemann matrices in Chapter 11.

Adrian met Lefschetz the year after Princeton elevated Lefschetz to full professor. He had ascended to the position by dint of sheer hard work and excellence, overcoming both social and physical handicaps.

Like Adrian's father, Lefschetz was born in Russia of Jewish parents. His family moved to France soon after his birth. He attended the École Centrale in Paris, where he studied engineering. Recognizing that his opportunities in France were limited because he was not a French citizen, he decided to leave the country.

In November 1905, around the time of Adrian's birth, Lefschetz came to the United States. While working as an engineer for Westinghouse in 1907, he lost both of his hands and forearms in a transformer explosion.

The handicap caused by his tragic and terribly painful accident effectively barred him from a career in engineering. However, there was one silver lining—Lefschetz turned to the pursuit of mathematics, his true calling. He enrolled in graduate school at Clark University in Worcester, Massachusetts and received his doctorate there.

He went on to teach at two midwestern schools, the University of Nebraska and afterwards, the University of Kansas. Lefschetz observed that both of these schools shared a common failing of state institutions in America—the requirement that they accept hordes of poorly prepared students whose only credential was a diploma from an accredited high school.[14] Neither institution had the prestige or intellectual stimulation of a top research university.

He finally received a call to come to Princeton in 1923 after his published research had gained him a national reputation. When he arrived at Princeton, the newcomer was virtually shunned by other faculty members. Not only was he the lone Jewish member of the mathematics department at the time, he cut an odd figure with his two artificial hands over which he wore shiny black gloves. Moreover, life's hard knocks had left their mark on his personality, leaving him toughened and outspoken. The students at Princeton epitomized him in this verse:

> Here's to Lefschetz, Solomon L.
> Irrepressible as hell.
> When he's at last beneath the sod
> He'll then begin to heckle God.[15]

Lefschetz gradually gained acceptance and admiration on campus. His students judged him to be a dynamic teacher who never resented an interruption and was open to students' suggestions.[16] In the words of one former Princeton student, "he was a sort of 'papa daddy' to graduate students."[17] Students gave him the moniker "G. W. F." (for "Great White Father").[18]

To Adrian, the nickname could not have been more apropos. He had lost his father less than two years before meeting Lefschetz. Lefschetz, married but childless, was twenty-one years his senior. The two men bonded and Lefschetz became one of his closest confidantes in matters both mathematical and personal.

After completing his postdoctoral fellowship year at Princeton, Adrian secured a position as Instructor in Mathematics at Columbia University, earning the princely salary of $3,000 per annum. He warmed early to the role of educator. The new instructor saw the science of mathematics as a sea of largely uncharted waters. As a fisherman, he understood the value of teaching others to maneuver in those waters.

The Alberts welcomed their firstborn during his second year at Columbia. In accordance with family tradition, they named him Alan (Elijah) after Adrian's father and ancient ancestor.

It was a sorrowful time for the nation. The economy was in shambles. Tensions were mounting in Europe. For Adrian, it was a critical time, as he now had a family to support in the midst of the Great Depression.

He longed for an offer to teach at his alma mater. It was not just that Chicago was his hometown. The University of Chicago held a special appeal to the budding algebraist as one of an elite group of universities in the United States that fostered research. Moreover, Dickson, one of the world's foremost authorities in Adrian's specialty, remained on the faculty there.

Adrian was not optimistic about the prospect. Despite the terms of the University's charter banning religious bigotry in hiring decisions, there were no Jews on the permanent mathematics faculty at Chicago. During the thirties, young Jewish mathematicians became aware that anti-Semitism prevailed throughout the nation's top research universities.[19] His hopes were deflated when Dickson himself sent a letter advising him to remain at Columbia.

But the young hopeful would not be defeated by bigotry. In the two years following his graduation, he churned out 21 research manuscripts. All of them found their way into leading mathematical journals.

The quality of his mathematics trumped any perceived discrimination at Chicago. After 30 years on the faculty, Dickson was approaching retirement. Chairman Bliss recognized that there would soon be a gap in the Department, since none of the younger faculty members could fill Dickson's shoes.

Bliss wrote to the Dean that he was interested in hiring Adrian Albert, explaining, "He is a very pleasant young man, Jewish in race...."[20] Bliss had done his homework, having already determined from Adrian's supervisor at Columbia that "they thought highly of his work in the classroom."[21] As for the candidate's qualifications, Bliss pointed out, "He was one of the ablest students we have ever had in our Department....His scientific activity has been unusual as the attached bibliography shows....Dickson knows his work and approves it highly."[22]

Having made the case for hiring his former student, Bliss got down to brass tacks. He had done his homework on that score as well. "I am proposing that we should offer him an assistant professorship at $3,500 and hope that he will take it. If he should not do so I think we should be ready to increase the offer to $4,000. Even this may not tempt him but it is likely to do so as both his own and his wife's family live in Chicago...."[23]

A bridge player, Adrian knew the rules of the bidding game. He informed the Department that Columbia had offered him an assistant professorship with a salary of $4,000 per annum and a substantial bonus for teaching during the summer session, bringing his prospective salary up to $4,666.[24] To buttress his argument he noted, "The prospects for promotion are very bright here." He countered that, if offered $4,000, "I should *definitely* accept it."[25]

On New Year's Day, 1931, he received a "Holiday Greeting" from Western Union with the message: "SUBJECT TO FORMAL APPROVAL OF BOARD OF TRUSTEES AM GLAD TO OFFER ASSISTANT PROFESSORSHIP JULY FIRST SALARY FOUR THOUSAND". He was ecstatic.

As soon as he learned of the appointment, Dickson could not wait to get his protégé inducted into the faculty's Quadrangle Club to add to his cadre of fellow bridge players. Despite the warm welcome he received upon joining the faculty, Adrian was something of an anomaly as the Mathematics Department's first Jewish member.

The faculty wives did their best to make Frieda feel welcome. Notwithstanding her light complexion and European parentage, they regarded her as someone of middle-eastern descent. To entertain her, faculty wives invited Frieda to tour the Egyptian mummy exhibits at the University's Oriental Institute. She was taken to the museum so many times she became sick of the place.[26]

The newly minted Assistant Professor would teach in the University's very first mathematics building, Eckhart Hall, then scarcely a year old. The mathematics building was the last of the original group of physical sciences halls to rise on campus.

Kent Chemical Laboratory was completed in 1893, one year after the University opened. The Ryerson physics building followed in 1894, predating Eckhart Hall by 36 years.[27] Yet another chemistry laboratory, Jones, opened its doors in 1929, a year ahead of Eckhart.

After physics, astronomy stood next in line. Due to the heavy industrial smog that often obscured Chicago's skyline, astronomy had to be located elsewhere. As mentioned earlier, traction king and former World's Fair board member, Charles T. Yerkes, agreed to build an astronomical observatory for the University's astronomy program. In 1897, the University's new observatory rose in Lake Geneva, Wisconsin.

Geology was the University's next priority. A son of German Jewish immigrants made that project a reality. Julius Rosenwald, who had revitalized the failing Sears, Roebuck Company, celebrated his fiftieth birthday by donating nearly three quarters of a million dollars to various causes that interested him. Helping to build the new Chicago campus was one of those causes. In 1915, construction workers topped off Julius Rosenwald Hall, which would house the studies of geology and geography.[28]

Because no mathematics building existed on campus during Adrian's schooldays, mathematics students were relegated to attending classes in Ryerson.[29] That pattern would change after Bernard Albert Eckhart agreed to donate funds for the project.[30]

Eckhart was born in Alsace, France in 1852, making him a contemporary of Adrian's father. Eckhart arrived in the United States during his infancy. He graduated from college in Milwaukee, Wisconsin and co-founded a milling company that later became known as B. A. Eckhart Milling Company.[31]

The upwardly mobile industrialist was not content to restrict his activities to manufacturing. A mover and shaker of his time, Eckhart became a trustee of such major enterprises as Continental Illinois Bank and Trust Company, Harris Trust and Savings Bank, Chicago Title and Trust Company, Erie Railroad Company, Armour and Company, Montgomery Ward and Company, and the city's Shedd Aquarium.

Eckhart deserves some of the credit for a massive engineering project critical to the city's future. He drew up the state law creating the Chicago Sanitary District that promised to end pollution along Chicago's shoreline and facilitate creation of a modern transportation waterway by joining Lake Michigan with the Illinois River.[32] He served as President of the Chicago Sanitary District's Board of Trustees from 1896 through 1900.[33] The city's exceptionally pure drinking water owes its origins to the water purification system developed by the Sanitary District.

Public service was a significant feature of his life. He served as a member of the Illinois Senate from 1887 through 1889. He also became a member of the Executive Committee on Capital Issues for the Federal Reserve Bank of the 7th District.[34] In 1910, Eckhart traveled to the International Congress on Education held in Vienna, Austria as the United States delegate.

The gift of the University's first mathematics building was one of the man's final contributions to his adopted homeland. Eckhart attended the dedication of his namesake along with his son, a University of Chicago graduate. The beautiful Gothic Revival structure was a fitting tribute to the civic-minded entrepreneur. A year after classes began at Eckhart Hall, death claimed the man from Alsace who had lived the American dream.

Eckhart Hall opened its doors in 1930. The structure was elegant, but the furniture inside was hideous. A note went to the administration demanding that it be returned and admonishing those responsible that the University's Purchasing Office should never again be permitted to select furniture. Instead, all furniture should henceforth be selected by "a woman of taste."

The new mathematics building adjoined the Ryerson Physical Laboratory. On the main floor, Eckhart Hall was separated from Ryerson by a graceful arch. But upstairs, mathematics flowed into physics seamlessly. Many years later, a former head of Chicago's

Mathematics Department attributed the spontaneity and intensity of mathematical life at Chicago to the concentration of all departmental activities at Eckhart Hall.[35]

Like his mentor, L. E. Dickson, Adrian was a bit of a showman, albeit in a more subdued manner. On the first day of class, he entertained a new batch of students by writing "A. Adrian Albert" on the blackboard with both hands at once, starting in the middle. His students affectionately dubbed him "A Cubed."

His salary would remain fixed at $4,000 for the next several years. Nonetheless, considering the state of the economy, he was one of the lucky ones. Although the dire state of the nation's economy during the 1930s undoubtedly depressed the salaries that the U. of C. was able to offer its professors, he felt fortunate to have a job and some income. I. N. Herstein wrote of his colleague, "Shortly after he got his Ph.D., the great economic depression started. Sensitive as he was to the suffering of others, he deeply felt the economic hardship that so many of his friends were undergoing."[36]

Years later, Adrian recounted some of his Depression-era experiences in an address to graduating students. He recalled that, during the Depression, students who were fortunate enough to complete their Ph.D. in mathematics dreaded the completion of their studies because "[t]here were almost no academic posts open, and the shock of being unable to find a post after years of study, and the award of a doctorate, did wreck the careers of some really talented people."[37]

Watching the suffering of his fellow scientists instilled a secondary purpose into his career—the desire to find a cure for the economic anemia afflicting his profession. He donated his time generously whenever asked to serve on a board or a committee, developing contacts among policymakers. In the process, he became adept at seeking out government and private grant programs to support the work of his fellow researchers. Over the years, his contacts in Washington and the private sector generated a continuous flow of funds into the University of Chicago's mathematics and physical sciences programs.

Less than 2 years after moving back to the Midwest, the 27-year old Assistant Professor learned of a unique opportunity on the east coast. He had a chance to join some of the world's top mathematicians for the opening year of a revolutionary new research institute—the Institute for Advanced Study.

The Institute's concept was a noble one—to allow some of the world's great thinkers time to learn and create, freed from academic duties. Private donors founded the Institute in 1930, the year after the stock market collapse, but it took three years to get the institution off the ground.

Pending construction of facilities for the Institute's School of Mathematics, the School moved into Princeton University's new mathematics building, Fine Hall. To Adrian, the lack of a separate building was not a drawback. The blending of the two faculties could only enhance the interaction between members of the mathematical microcosm.

The notion of an entire academic year with the freedom to devote full time to research was irresistible. The University of Chicago agreed to grant him a leave of absence for the 1933–34 academic year and pick up half of his salary to boot. The Alberts would reside in a familiar setting near Princeton University where the young couple had spent his postdoctoral year. They were about to embark upon a seminal interlude in their lives.

The Alberts had purchased a Willys sedan the previous year for $495 with a hundred dollars down. They packed up their belongings and drove to Princeton with their two-year old son in tow.

Another member scheduled to arrive for the Institute's opening year was Albert Einstein. While Adrian, as a younger scientist, had a one-year appointment, Einstein would come as one of five members of the Institute's first permanent faculty, along with Oswald Veblen, Hermann Weyl, John von Neumann, and James Alexander.[38]

Adrian and Frieda had agreed to greet Einstein and his wife Elsa upon their arrival at the railroad station and drive them to campus. Shortly after the Einsteins climbed into the back seat of the Alberts' Willys, the heavens opened up. Bullets of rain began to pummel the roof of the sedan mercilessly. Frieda turned around to converse with the Einsteins and stopped in midsentence. She was mortified to see the downpour invading the Willys and dripping onto the great man's homburg. Whether Einstein was being gracious or was simply oblivious, she could not tell. But Einstein paid no attention to the incident and went on with their conversation as if nothing had happened.[39]

A confluence of events made Princeton a magical place for a young mathematician in 1933. A flood of Europe's best and brightest mathematical émigrés wound up in Princeton. The opening of the new Institute provided an opportune haven for some of these refugees. To the refugees, the Institute recreated the aura of the recently constructed Mathematical Institute at Göttingen.

Princeton University's beautiful mathematics building, completed in 1931, materialized in time to host the illustrious newcomers. Fine Hall was a perfect setting in which to replicate Europe's teatime tradition. Afternoon teas offered a rare opportunity for young mathematicians like Adrian to rub shoulders and share ideas with many of the world's finest mathematical minds. Between Princeton's faculty and the Institute's permanent and temporary members, Fine Hall fairly sizzled with mathematical talent.

Adrian joined in enthusiastically. In response to Chairman Bliss' demand for a report of his activities, he replied, "The principal change in my Chicago interests has been the substitution of chess for billiards. We have a lovely common room here, and have tea at 4:30 daily."[40]

He was like a kid in a candy store. Surrounded by scholars he admired, Adrian wanted to glean as much as possible from as many of them as possible. He attended a series of lectures delivered by Hermann Weyl‡ as well as a course of weekly lectures given by Emmy Noether, who shuttled between the Institute and Bryn Mawr College in Pennsylvania where she held a temporary teaching position after emigrating from Germany in the fall of 1933.

Emmy Noether likewise came to hear Adrian lecture and wrote to Helmut Hasse at the Mathematical Institute in Göttingen, "Albert...lectured before Christmas something hypercomplex after Dickson, together with his Riemann matrices."[41] She complained to Hasse that her American audience was unaccustomed to her German-style abstract algebra, commenting, "I have essentially only research fellows in my audience, besides of Albert and Vandiver, but I have to be very careful because they are used to explicit computations, and some of them I have already driven away!"[42]

Noether's lectures at the Institute touched on algebraic number theory. Adrian's interest in her lectures was far from academic. He had only recently collaborated with the German team of Hasse, Noether,

---

‡ More about the impact of Weyl's lectures in Chapter 12. Also see Chapter 13 for discussion of the exodus of many of Europe's scholars.

and a third mathematician, Richard Brauer, in proving a groundbreaking theorem on the theory of algebras.‖ Their theorem was already finding applications in class field theory, a developing aspect of number theory. During his stay at the Institute, Adrian became fascinated with algebraic number theory and read volumes on the topic, believing that he was on the track of a new theory.[43]

Adrian met Brauer in person for the first time at the Institute. Brauer arrived in the United States in November 1933 after dismissal from his teaching position in Königsberg due to Nazi legislation removing all Jewish university teachers from their posts. Brauer passed through Princeton on his way to Kentucky, where he had secured a temporary post.[44] Adrian took the bewildered refugee under his wing and brought him around to meet the great Wedderburn.[45]

Despite the Institute's stated policy of affording scientists an opportunity to conduct research unfettered by academic duties, Institute members were expected to share their research. Adrian taught a yearlong seminar on algebra at the Institute. Meanwhile, he commenced a dizzying flurry of research and writing.

He turned his attention to solving a problem involving an "algebra of quantum mechanics" posed to him by the stellar trio, John von Neumann, Eugene Wigner, and Pascual Jordan.# He published papers on a variety of topics, such as "modular fields," "cyclic fields," "Riemann matrices," and "nonassociative algebra."[46]

While teaching and conducting his own research, he took on two new jobs: serving as Associate Editor to Lefschetz for Princeton University's *Annals of Mathematics* and helping Wedderburn finish a new book that needed extensive referencing. Ordinarily, faculty members edited the *Annals* in-house at Princeton.[47] Adrian would continue as Associate Editor for a decade although he lived in Chicago most of that time.

In the process of lending a hand to Wedderburn, Adrian became acquainted with Wedderburn's brilliant 23-year old doctoral student, Nathan Jacobson, already hard at work on the book when he arrived. Wedderburn's two protégés sensed a bond of commonality from the start.

‖ Chapter 10 tells the controversial story of their proof.
More on the algebra of quantum mechanics in Chapters 12 and 16.

Jacobson and Brauer were but two of the mathematical friends Adrian acquired in Princeton. He also became friendly with Lefschetz' protégé, Oscar Zariski. Zariski, born in Belarus, had fled the ravages of World War I with his parents as a child. He eventually earned his doctorate in Rome, but then began to suffer from the Fascist, anti-Semitic regime of Mussolini. With Lefschetz' help, Zariski managed to immigrate to America in 1927.[48]

Zariski taught at Johns Hopkins University between 1927 and 1937, but often visited Lefschetz at Princeton during that period. Zariski increasingly gravitated to Lefschetz as he recognized that Lefschetz' methods were altering the field of algebraic geometry.

Adrian and Zariski compared notes concerning their ancestry. Zariski believed he could trace his lineage back to that ancient sage, the Vilna Gaon. The two men concluded that they were most likely distant cousins.

During the Alberts' stay in Princeton, a letter arrived from the American Association for the Advancement of Science. The AAAS had nominated Adrian for "fellowship" in the Association, a singular honor limited to those who have "advanced science by research."[49]

Upon returning to Chicago, he made up for lost time by teaching two courses during the summer of 1934. He dropped a note to Jacobson, confiding that he needed to recuperate after the strenuous year in Princeton. He urged Jacobson to come to Chicago so they could continue working together.[50]

Chapter 5

Kinships and Kindred Spirits

Three years after joining the faculty at Chicago, A.A. Albert celebrated the graduation of his first doctoral student, a woman named Antoinette ("Tony") Killen Huston. Ironically, his first "mathematical child" was a fellow Phi Beta Kappa member of his 1926 college graduating class.[1] She had worked her way through graduate school by serving as secretary to Chairman Bliss.[2] In the process, Tony Killen met and married a graduate student named Ralph Huston and started a family.

Adrian's students recalled, "[n]obody in his classes didn't like him."[3] He urged them not to give up and tried to simplify things for them.[4] The soft-spoken teacher had an "open door" policy—students were free to drop by his office any time they had a question. He was reputed to be an easy grader—most of his students received As, and it was rumored that no one received any grade lower than C.[5] On the other hand, when it came to the ultimate prize—a doctorate in mathematics from the University of Chicago—the bar was high. Like his faculty brethren, Adrian avoided the temptation to help his students solve the problems assigned for their dissertations.[6] It was bitter medicine, but puzzling things out for themselves made them better mathematicians in the long run.[7]

Many doctoral students in mathematics at Chicago faltered and fell by the wayside. Some of Adrian's fellow faculty members took a "tough guy" approach—"Don't bother me until you've proved the theorem." He had a reputation for taking better care of his students.[8]

He also provided an indispensable form of assistance to needy students. On occasion, he was known to buttonhole a student in the hallway and stick a grant or fellowship application in the student's hand with the advisement, "Here—fill this out and send it in." Afterwards, he might lobby the funding agency to assure that the amount of the student's stipend was adequate.[9]

The Alberts' middle child, Roy, was born in 1935. Within a few years, I came along. Tony Huston wrote, "I positively suffered while opening the announcement. I was so afraid Nancy Elizabeth might be another Mr. Albert....I want to find out just how one guarantees that the third baby will be a girl."[10]

Adrian's wish to have Nathan Jacobson join him at Chicago was granted in 1936 when Jacobson received a National Research Council Fellowship to come and work with him.[11] The two men, both disciples of Wedderburn, shared similar interests in their research. During the year, their friendship grew into a warm mathematical kinship.

The thirties were stressful times for Jacobson, who bounced from one university to another in search of a permanent teaching position. The promising young mathematician, whose family had fled discrimination in their native Poland when he was a child, revisited anti-Semitism at America's top universities. He found that "[t]here were no Jews on the mathematics faculties of Yale or Harvard, one at Chicago, and three at Princeton, and it had been decided either by the department or the administration that these two departments could hire no more Jews."[12] After his year at Chicago, Jacobson succeeded in securing a teaching position at the University of North Carolina.

In the summer of 1938, Adrian organized a Conference on Algebra at Chicago. The Conference kicked off a summer long program devoted to concentration in algebra.[13] Jacobson much later recalled, "I was invited to participate [in the Conference] and to be a visiting lecturer for the summer quarter. As such I gave a course entitled *Continuous Groups* that was perhaps more notable for its audience, that included Irving Kaplansky and George Whitehead, than for its content...."[14]

At the time, Kaplansky was fresh out of college with a bachelor's degree from the University of Toronto, where he was influenced by his professor, Richard Brauer.[15] During his last year of college, he had demonstrated his potential by winning the first-ever William Lowell Putnam Mathematical Competition for undergraduate students. The competition came to be regarded as the high water mark of intellectual achievement for aspiring mathematicians.

Besides Jacobson's course, Kaplansky attended a class of Adrian's that summer.[16] Years afterwards he remembered, "The summer long event at Chicago in 1938 was an algebra program dominated by Adrian.... That summer put a stamp on me that lasted a lifetime."[17]

Chicago was becoming a Mecca for mathematicians as Adrian brought many of the world's most brilliant mathematical minds to campus. Saunders Mac Lane, a younger mathematician spending the year at Chicago after an instructorship at Cornell, assisted Adrian in organizing the Conference. Mac Lane had received his master's degree at the University of Chicago but traveled to Göttingen for his doctoral studies.

Their goal in designing the Conference was to introduce concepts such as "rings," "function theory," "field theory," "structure theory for associative algebras," and the "theory of integers for rational division algebras" to a new audience, and to show how those concepts were interconnected.[18]

Fourteen top mathematicians graced the podium. Dickson kicked off the event with a lecture on the history of "quaternions" and other "linear algebras." According to the journal *Science*, "Professor Dickson described the long and difficult search for these numbers and their final discovery." *Science* also noted that "Professor A. A. Albert, of the University of Chicago, gave a new proof for some properties of the algebras built up by combinations of cyclic algebras...."[19]

Kaplansky's former professor was one of the featured speakers. Brauer spoke on the structure of "division algebras," while Jacobson lectured on "Lie algebras." Jacobson explained that "one example of such algebras is furnished by ordinary vectors, using vector multiplication." Emil Artin, a leading German mathematician, introduced the audience to the theory of "quadratic forms."

Mac Lane lectured on "field theory," while Lefschetz and Zariski expounded on "algebraic geometry." Lefschetz caused quite a controversy when he declared that many results in the geometry of curves and surfaces previously found by analytical methods could be obtained by simple algebraic devices.

The Conference helped to spread the word about new frontiers in mathematics and shaped the thinking of the next generation of American algebraists. The gathering made such an impact that historians of mathematics marked it as a significant event.[20]

The event was accompanied by social get-togethers where spouses were included. Brauer brought his wife Ilse, a former classmate in mathematics at the University of Berlin. Ilse was a round, motherly woman whose warm smile accompanied her keen intellect. The meeting marked the beginning of a strong friendship between the Alberts and the Brauers.[21]

When Adrian organized a second algebra conference in 1941, he invited Jacobson for an encore performance. Seated in the audience at Jacobson's lecture was a member of a rare breed—a female student with a master's degree in mathematics. The student was Florence ("Florie") Dorfman, whose doctoral supervisor was none other than Jacobson's close friend, Adrian Albert. A natural attraction developed between the bright, impressionable student and the handsome, inspiring lecturer from North Carolina.

Soon after the United States entered World War II, the Navy chose the University of North Carolina as the site of a Pre-Flight School for training Navy pilots. In 1942, Jacobson and other potential Pre-Flight School teachers were sent to the University of Chicago for a month long teacher-training course.[22]

That serendipity brought Jacobson and Dorfman together again and the pair became inseparable. Within a short time, Jacobson had asked Florence to marry him. It would mean giving up her doctoral studies at Chicago, since he had secured a position as Visiting Associate Professor at Johns Hopkins.

Florence could not refuse the brilliant, articulate, and kind-hearted man who shared her love of mathematics. The Alberts joined the festivities as guests at the couple's wedding. Florence assumed several new roles—mathematical wife, sounding board, and occasional coauthor with her husband.

Besides his genetic family and mathematical kin, Adrian loved his alma mater, and especially loved its Mathematics Department. Eckhart Hall would remain his second home for forty years. He confided to Frieda, "Wherever I go I see that our Department is a paradise compared to all but perhaps Princeton, the Institute, and Harvard."[23]

His love for the University of Chicago was due in no small part to the vision of its first President, William Rainey Harper. When Harper accepted the presidency in 1891, it was with the understanding that he would have a free hand in developing the institution along the broad lines he wanted. Under Harper's plan, "the University of Chicago would include an undergraduate college, but senior professors would be freed from heavy teaching loads in order to pursue research."[24] It was the freedom to pursue research that led to Adrian Albert's productivity and his loyalty to the University.

PART TWO

Framework for a Mathematical Mind

Chapter 6

Mathematese and Mathemagicians

The techniques that A.A. Albert and others introduced at the 1938 Conference on Algebra were new and somewhat confusing even to experienced mathematicians. The abstract science had drifted further than ever from the reach of the average person.

Mathematics tends to befuddle the uninitiated on account of "mathematese" and "mathemagicians." Mathematese is a language full of teases. Mathemagicians, a/k/a mathematicians, make things appear to disappear.

The ancient Greek mathematician Pythagoras (569?-500? B.C.) and his followers cloaked their discoveries in mystery and secrecy. Using their secret methods, they managed to divine the revolutionary idea that the earth is a sphere revolving around a central fire. The Pythagorean brotherhood believed that numbers rule the universe. They were some of the first scholars to discern the inner workings of nature through mathematics.

A controversy roils among mathematicians that may never be settled. Because no one has ever been able to touch or see a number, some believe that mathematicians dreamed up the notion of numbers in their imaginations. Others are equally insistent that numbers derive directly from nature. In that case, mathematicians are merely deciphering nature's code when they investigate the science of numbers.

Did mathematicians create mathematics, or is mathematics a set of inherent laws governing forces of the universe such as planetary motion, gravity, electromagnetism, atomic energy, and perhaps life itself?

One thing is certain: mathematicians invented the language and methodology of their craft. For starters, they convinced most people that numbers were "real." When they believed a mathemagical concept might be particularly useful, they conjured it up by calling it an "object."

Mathematicians wanted to perform tricks with their numbers, so they divided them up into "sets," and even created an "empty" set consisting of absolutely nothing. They struggled with numbers that refused to behave such as "negative numbers" and "irrational numbers."

In their mind's eye, the mathematicians could "see" all of the "real numbers" lined up like calibrations along a thermometer with zero holding the most respected spot in the center. Their imaginary thermometer extended all the way to infinity. To the right of zero stood all of the "positive numbers" they could dream of, including whole numbers, fractions, decimals, square roots, and even the elusive π. To the left of zero stood all of the "negative numbers" like the frigid minus twenty, also known as twenty below zero.

Because irrational numbers like π and the square root of 2 refused to behave like the so-called "rational numbers" that can be depicted by fractions, they were disrespected like the stepchild Cinderella. They gained respectability once mathematicians learned that irrational numbers could be depicted by a chain of unending and nonrepeating decimals.

Mathematicians believed that things were going quite well until some of them discovered a number even more abstract than the numbers they had imagined thus far. The new number was the square root of minus one.

This was a surprising development—even mathematicians were perplexed by the new number. Gottfried Wilhelm Leibniz (1646–1716) waxed lyrical about the subject. He declared, "The Divine Spirit found a sublime outlet in that wonder of analysis, that portent of the ideal world, that amphibian between being and not-being, which we call the imaginary [square] root of negative unity."[1] Although no one had ever proved the existence of *any* so-called real number, mathematicians decided to call their new find an "imaginary number." They dubbed the new number "i." This abbreviated notation allowed them to combine their enigmatic imaginary numbers with real numbers in "complex numbers" so seamlessly that imaginary numbers became almost invisible.

In order to manipulate their sets of numbers, these seers defined delicate "operations" like addition and its inverse, subtraction. Next, they defined multiplication and its flip side, division.

By applying an operation such as multiplication to one set of numbers, they found that they could create a second set of numbers. For example, by multiplying each of the members of the set "1,2,3,4" by 2, they could create the set "2,4,6,8". From such operations came the concept of a "function," which "describes how one set of quantities depends on or varies with" another set.[2]

The seers began to visualize ethereal universes, or "systems," where they could subject number sets to various operations. Since mathematicians defined the "systems," they could choose to include or exclude whatever they pleased in each system.

One of the systems they defined was called a "group." Counterintuitively, the term "group" does not imply that the system contains more than one set of elements. What makes it a group is that it includes both a set of elements and one operation which is used to combine any two objects in the set (a "binary operation"). Groups are also characterized by a strict set of rules.

Mathematicians also defined magical "rings" which were groups where two operations, addition and multiplication, could be performed on any two objects in a set. Again, the all-important rules governed, and imposed a structure on the otherwise lawless sets. They called the rules "axioms," from the Greek word meaning "something worthy."

Not all rings were alike. They constructed special rings called "fields" which, unlike ordinary rings, permitted division.

Since the real and imaginary numbers they manipulated were often quite unwieldy, mathematicians needed a magical trick to simplify them. So they invented a new science and named it "algebra". They borrowed the name "algebra" from the Arabic phrase, "al jabr w'al muquabalah," meaning "reunion and opposition"—two manipulations needed for balancing the yin and yang of equation problems.[3] Equations would become the keys to unlocking the mysteries of mathematics through algebra.

The ultimate "tease" in mathema-tease involves the very term, "algebra." To further befuddle any novices seeking to unlock the secrets of their trade, mathematicians gave the term two distinctly different meanings. "Algebra" is a branch of mathematics concerned with equations and arithmetic operations such as addition, subtraction, multiplication, and division. But "algebras" are a class of rings having an additional vector space structure. "Algebras" permit multiplication of two "vectors" to form a third.

To invent algebras, they had to engage in a bit of alchemy, since they were essentially transforming "vectors," which belonged to the Realms of Astronomy and Physics, into an ethereal mathematical concept. After all, vectors represented forces having magnitude and direction.

Nineteenth century mathematicians conceived of vectors as directed segments of unit length residing in mini-universes they called "vector spaces." Vector spaces are systems that include all of the vectors within them, just as outer space includes all of the planets and stars. Physicists had always depicted their vectors as arrows to show that they represented both magnitude and direction.

In order to manipulate their algebras, mathematicians imposed a third operation which they called "scalar multiplication." They visualized "scalars" as quantities or points on a scale. In physics, scalars were numbers representing speed, time, or temperature.

By erasing any semblance of the real world in their computations, they could employ the same computation to represent a variety of phenomena. They replaced the real world notion of scalars representing speed, time or temperature by the abstract notion of scalars as numbers in a field. To describe the extra operation, mathematicians created the shorthand term "over a field" in lieu of the bulky phraseology "multiply by the numbers (scalars) in a field." Every algebra has a base field. The base field can be the field of real numbers or some other field.

William Hamilton (1805–1865), an Irish professor of astronomy with a background in mathematics, discovered an algebra that future mathematicians would rave about. As an astronomer, he investigated the mysteries of space, time, forces, and motion.

Hamilton wanted to compute rotations in space. He chose vectors as a means of depicting the various segments of a celestial body's orbit.

For his first magic trick, Hamilton decided to avoid using conventional geometry in his computations. Instead of a spacial (geometrical) representation, he used numbers and symbols. He converted his lines, points, and angles into real numbers and imaginary numbers such as the imaginary root of minus one, that is,

"i". The mystical i was an essential ingredient in his computations, since he knew that "multiplying by i gives a 90 degree counterclockwise rotation" to a vector.[4]

Using the streamlined approach advanced his research, but he was left with computations that might be as unwieldy as $(3 + 8i + 2i^2 + 6i^3)$ times $(5 + 2i)$. He needed to transform that convoluted mélange of numbers into something he could work with. Like most mathematicians, he used equations as the medium for boiling that heady brew down into its most elemental form. Remembering that those deceptively simple-looking "i"s actually represented the square root of minus one, he was able to transform $2i^2$ into minus 2, and $6i^3$ into minus $6i$. So that $(3 + 8i + 2i^2 + 6i^3)(5 +2i) = (3 + 8i -2 -6i)(5 + 2i)$.[5] He proceeded to use algebraic equations to further reduce his number soup all the way down to "$1 + 12i$". Hamilton's result, "$1 + 12i$", was considered a complex number because it contained both real and imaginary numbers. (Mathematicians would later come to view the complex numbers as 2-dimensional algebras.[6])

For his next feat, using a clever sleight of hand, Hamilton made the numbers disappear entirely. He transformed the specific numbers, 1 and 12, into "generalizations" of numbers by renaming them "a" and "b". So now, "$1 + 12i$" became "$a + bi$." So far, he had transformed a maze of numbers into three letters connected by the mystical sign of the cross. But Hamilton was not yet satisfied.

Now that Hamilton had boiled his complex numbers down into the compact "$a + bi$" symbols, he vaporized the last vestige of his numerical mumbo jumbo. In an ultimate feat of mathema-tease, he decreed that, henceforth, he would call the complex number "$a + bi$" the "ordered pair" of numbers, (a,b). Hamilton had magically transformed his two-dimensional vectors into the compact couple (a,b).*

Hamilton's recasting of the complex numbers into couples made it easier for him to add and multiply his two-dimensional vectors. This solution would have sufficed for calculating the paths of celestial bodies moving in two-dimensional space. If that were so, he could rely on the rule of arithmetic that told him it made no difference which number he wrote first—the result would be the same.

---

* See Chapter 7 for a discussion of Descartes, Gauss, and the Cartesian plane that Hamilton had studied before developing his innovative representation of the complex numbers.

However, in reality, the rotations of celestial bodies were not two-dimensional. They moved in an elliptical fashion in three-dimensional outer space. And when Hamilton attempted to multiply three-dimensional vectors, the system broke down. He kept coming up with zero, and he was not seeking the "empty set."

He was trapped like Houdini in a straitjacket. Hamilton was constrained by the earthly axiom of arithmetic stipulating that if two numbers are multiplied together, the order of multiplication is immaterial. For example, (2 x 3) = (3 x 2). This was the "commutative law." The commutative law stood in Hamilton's way when he tried to multiply three-dimensional vectors.

Then something unexpected happened. It was a magical revelation. Hamilton could not tell where it came from, since it happened inside his head. He broke open the straitjacket by freeing himself from the rule of commutativity. He surmised that the order in which one calculates rotations in three-dimensional space affects the outcome. By suspending the age-old commutative law, Hamilton found that he could not only multiply three-dimensional vectors, but four-dimensional vectors as well. He named his new four-dimensional objects "quaternions." Since they were twice as complicated as the complex numbers, he decided that they were "hypercomplex numbers."

Hamilton immediately recognized the importance of his discovery. He considered the quaternions "his masterpiece and his title to immortality."[7] By virtue of this groundbreaking discovery, Hamilton paved the way for the dramatic changes that swept over algebra during the twentieth century. He also paved the way for a new paradigm of the universe where three-dimensional space took on a fourth dimension—time. Hamilton's discovery of a new approach to formulating the laws of motion led to adoption of the term "Hamiltonian dynamics."

Chapter 7

From Al Jabr to Abstract Algebra

A high school student confronted with an algebra text for the first time is apt to ask, "Why should I care about algebra? What use will it ever be to me?"

Algebra is an efficient science. For the cost of paper and pencil, algebra can simulate a variety of phenomena, both natural and manmade. The beauty of algebra is that its abstract methods can reveal answers to a wide range of questions. It can direct missiles to their targets or guide spaceships safely to the red planet.

In a sense, algebra is an extension of arithmetic, since algebra involves reasoning about quantitative relationships.[1] These relationships are generally everyday arithmetic operations like addition, subtraction, multiplication, and division. If Johnny needs an extra nickel to buy a candy bar, there is a relationship between the nickel and what Johnny already has in his pocket. The operation involved in this relationship is addition, since Johnny has to add the nickel to his other change in order to afford the candy bar.

In its simplest form, algebra is a branch of mathematics where arithmetic relations are expressed by using letters or other symbols instead of numerals. Symbols provide a convenient mathematical shorthand allowing complicated concepts to be expressed concisely.[2] These symbols are combined in algebra to form equations. For example, $(a + b) = (c + d)$ can be used to represent $(1 + 4) = (3 + 2)$. The term "equation" is used because each of the expressions is "equal" to the other.

What differentiates algebra from arithmetic is that algebra concerns thinking about the general principles that apply to arithmetic. It is, in essence, a generalization of arithmetic.[3]

Despite its theoretical approach, algebra arose out of mankind's need to solve real world problems. The use of equations with unknown variables often referred to as "x" proved to be an effective device in finding these solutions.[4]

A.A. Albert expressed this "real world" view of algebra in his text, *College Algebra*. In it, he likened mathematical formulas to machine tools that operate on raw materials to yield finished products. "The raw materials of mathematics arise from the problems of the physical world and are usually numbers. The finished products are the solutions of problems."[5]

On the other hand, he recognized that, once one left the realm of college algebra and entered the world of "modern algebra," the subject became increasingly abstract. As examples of such abstractions, he cited "concepts of group, ring, integral domain, and field..."[6] These concepts form the basis of modern algebra. To Adrian, his fairyland of imaginary theorems was so real, he could live in it.

Adrian and his brethren in the mathematical community considered themselves researchers in "pure" abstract algebra, that is, the world of ideas. Yet, their theoretical research would eventually find many real world applications. The "matrix theory" that A.A. Albert explained in his intermediate algebra text became a key tool in the new cryptographic methods he developed during World War II.* The theory of "finite fields" that he presented in one of his advanced algebra texts formed the basis for digital computers and other devices using binary digits developed in the 1950s.[7]

The "linear algebra" that he spent a lifetime investigating is a key ingredient underlying quantum mechanics as well as today's cutting edge investigations into "string theory," the quest for one master equation to describe all matter and forces at work in the universe. Today's string theorists employ algebra in their efforts to reconcile Einstein's theories about the forces of gravity upon large objects, such as planets, with quantum theorists' findings about the very different behavior of forces in the subatomic world.

---

* See discussion of WW II codes in Chapter 13.

Although we tend to take the scientific advances wrought by algebra for granted, it had an incredibly long gestation period. The Babylonian civilization, *circa* 2000 B.C., was the first to invent algebra. However, the Babylonians failed to develop a system of mathematical symbols to express their craft. Nor did they commit their knowledge about the properties of numbers to writing.[8]

It was not until the middle of the third century A.D. that a Greek scholar named Diophantus introduced symbols for use in algebraic equations.[9] Although his methods originated in Babylonian mathematics, Diophantus has often been called the father of algebra because his work made a singular contribution to the science.

The abstract, or "modern," algebra practiced by pure mathematicians today differs dramatically from the rudimentary algebra employed by Diophantus. Modern algebra did not truly blossom until the twentieth century. Hamilton was one of the key innovators who laid the groundwork for moving the science to the level of abstraction it has reached today. The discoveries of Hamilton and a small band of like-minded mathematical ancestors provide a context for understanding the work of Adrian Albert and his contemporaries. The ladder of innovations constructed by these influential mathematicians led directly to the seemingly sudden surge of advances introduced by modern algebraists.

The roots of modern algebra can be traced to France, where mathematicians such as Descartes and Galois formulated some of the fundamental concepts that allowed the science to develop. René Descartes (1596–1650), an admirer of Galileo, believed that the study of mathematics would lead mankind to discovering the truth about how the universe works.

Like Galileo and Pythagoras, Descartes believed the sun to be the center of the universe. Despite Descartes' close ties to church leaders, Galileo's fate caused him to fear the scourge of religious bigotry. He was traumatized when he heard that the Inquisition had forced Galileo to recant his theories, and wisely postponed publication of his own manuscript because of its assumption that planets revolve around the sun.

Descartes decried civilization's rudimentary understanding of mathematics. Yet, he was hopeful that the "science known by the barbarous name Algebra, if only we could extricate it from that vast array of numbers and inexplicable figures by which it is overwhelmed,...might display the clearness and simplicity which...ought to exist in a genuine Mathematics."[10] Little did he know that, building on his own discoveries, twentieth century mathematicians would realize his dream of freeing algebra from the baggage that overwhelmed it.

He is credited with reducing the science of geometry to algebra, thereby creating the new science of "analytic geometry."[11] Geometry deals with graphic concepts like points, lines, angles, and surfaces.[12] Algebra's approach of dealing with letters or other symbols used to represent numbers and other mathematical entities would appear to be entirely different. Descartes' genius was in bringing the two concepts together.

His method established the correspondence between the "domain of segments," a geometrical concept, and the "field of real numbers," an algebraic concept. Descartes found that the use of letters for known and unknown variables was superior to geometry, since it enabled him to exhibit certain distinctions without the clumsy conventions of drawing.[13] At the same time, analytic geometry enabled Descartes to visualize algebraic formulae in a way that ordinary algebra did not.

Underlying his new science was a mapping of the "plane," a two-dimensional surface. In a letter to a friend, he wrote, "In the solution of a geometrical problem I take care, as far as possible, to use as lines of reference parallel lines or lines at right angles." E. T. Bell describes the basic idea: "Lay down any two intersecting lines on a plane....The whole plan will be laid out with respect to *one* avenue and *one* street, called the *axes*, which intersect in what is called the *origin*.... The avenue-number and street-number...enable us to fix definitely and uniquely the position of any *point* whatever with respect to the *axes*, by giving the *pair* of numbers which measure its *east* or *west* and its *north* or *south* from the *axes*; this pair of numbers is called the *coordinates* of the point...."[14]

Descartes did not invent the grid system often associated with his name,[15] but it is undisputed that he had been studying coordinate geometry for many years before publishing his famous work, *La Géométrie*. The mapping inherent in his work was critical to the development of modern algebra. Future mathematicians would view the entirety of points in a "Cartesian plane" as a vector space.

The implications of the grid system were wide-ranging. Not only could any point be described by the label of the grid where it lay, but any curve could likewise be characterized in terms of the sequence of grid cells it crossed.[16] As a result, a geometric curve could be identified with the algebraic connection between the horizontal axis x and the vertical axis y. This led to the theory of functions focusing on the connection between x and y for a given curve.[17] The grid system became a means of expressing, in a graphic way, the relationship between two quantities that are "functionally related" to one another, such as the relationship between the pressure in a gas and the size of its container.[18]

Descartes applied his new methods, computing curves by means of algebraic equations assuming a system of axes and coordinates, with stunning success. He made the tricky business of computing multiple curves simultaneously seem simple.

His contemporary, Girard Desargues (1591–1661), invented a new way of doing geometry that would come to influence future algebraists. Desargues, whose interests lay in practical applications for mathematics such as stone cutting, developed a theory about how triangles meet when observed in perspective. To explain his theory, he wrote an essay on the results of taking plane sections of a cone. Desargues' efforts led to the discovery of another new science, "projective geometry."[19] About 300 years later, Desargues' invention proved to have applications to quantum mechanics, the science that enabled physicists to split the atom.

Imaginary numbers were actually introduced, not by Leibniz, but by the Italian mathematician Girolamo Cardano (1501–1576) and an Italian engineer-architect named Rafael Bombelli (1526–1572) about 100 years before Leibniz rediscovered them.[20] Bombelli and Cardano could not smell or touch these numbers, but they wrote them down, since the numbers were needed in their equations. Still, mathematicians had a hard time deciding whether they should be considered legitimate numbers.[21]

Today, algebraists welcome these mystical creatures and accept their existence, although mathematicians continue to use the term "imaginary numbers." However, note that if you multiply the square root of minus one ("i") by itself, the result is the rather mundane "-1".

Nineteenth century mathematicians began to look at numbers as part of a "system" that involved arithmetic operations. In the view of mathematician and historian of mathematics, B. L. van der Waerden, modern algebra began with Evariste Galois (1811–1832). Galois was the first to introduce the concepts of fields and groups.[22] According to van der Waerden, "Before Galois, the efforts of algebraists were mainly directed to the solution of algebraic equations....Galois, on the other hand, was the first to investigate the *structure* of fields and groups, and he showed that these two structures are closely related."[23]

France was regarded as the epicenter of the mathematical world during the 18th and most of the 19th centuries. Political upheaval in Europe, especially unification of the German states under Prussia and the surrender of Paris to Prussian troops in 1871, seriously undermined French academic institutions. Meanwhile, the great Johann Karl Friedrich Gauss (1777–1855) and his disciples gradually brought about a shift in the balance of mathematical power. Gauss was born in the Duchy of Brunswick, which now belongs to Germany.[24] He spent nearly 40 years as director of the astronomical observatory at Göttingen, Germany.

To Gauss, mathematics, astronomy, and physics were inextricably linked. He made contributions to all three sciences.

Adapting the Cartesian grid system to fit his needs, Gauss designated the horizontal x-axis as the real number line and the vertical y-axis as the line of imaginary numbers. This enabled him to label points on the x-axis as real numbers and points on the y-axis as imaginary numbers. It then followed that any point in the plane could represent a complex number containing both real and imaginary numbers.[25] Hamilton, who was familiar with Gauss' geometric method of representing the complex numbers, used it as a frame of reference when he developed his own streamlined algebraic representation of the complex numbers—the numbers that were so essential to computing celestial rotations.[26]

Richard Dedekind (1831–1916) and Bernhard Riemann were two of Gauss' students. In 1879, Dedekind refined the definition of a "field." He discovered that the number sets in his fields obeyed the law of commutativity with respect to both addition and multiplication.[27]

Dedekind also investigated the theory of "prime numbers."[†] A "prime number" is "any number greater than one which has as its

---

[†] The theory of Prime Numbers belongs to the study of "number theory." While mathematicians revel in the mental acrobatics involved in number theory, E. T. Bell has observed that "the by-products of these apparently useless investigations amply repay those who undertake them by suggesting numerous powerful methods applicable to other fields of mathematics having direct contact with the physical universe." (E. T. Bell, *Men of Mathematics*, at p. 64.)

divisors (without remainder) only 1 and the number itself."[28] He struggled with the problem that, when prime numbers were combined with imaginaries in complex numbers, they refused to obey the "fundamental theorem of arithmetic." The theorem stated that any nonprime whole number greater than one can be written as the product of a unique set of prime numbers.[29] A solution was needed to make the formerly pristine prime numbers comply with this basic law of arithmetic.

Dedekind's predecessor, Ernst Eduard Kummer (1810–1893), had partially solved the problem by separating the complex numbers into prime factors. Kummer named his discovery "ideal numbers." Drawing on Kummer's work, Dedekind developed the concept of an "ideal," a generalization of the concept of "number."[30]

Another great German mathematician, David Hilbert (1862–1943), took up the study of Dedekind's ideals. Hilbert extended Dedekind's work on ideals in "algebraic number fields" and coined the term "ring," which would eventually become one of the most dominant topics of modern algebra.[31]

Hilbert and his disciples inspired a new way of thinking about numbers. Even the humble ordinary integers were no longer considered mere counting devices, but members of a ring endowed with certain inherent properties. "Ideals" were seen as subsets of rings. Mathematicians increasingly focused more on the *properties* of number systems than on the numbers themselves.

Mathematicians are uncertain about the precise derivation of the term "ring." A likely explanation is that mathematical rings are "closed" like a circle.[32] In mathematical jargon, rings are "closed" in the sense that, if the operation of addition or multiplication is performed on any two members of a ring, the result is always another member of the ring.

The misleading part about the term is that rings are not required to be round. Since rings were invented in the minds of their mathematical beholders, mathematicians can choose to visualize them as they wish. For example, although some mathematicians visualize fields as large flat Cartesian planes filled with numbers, fields are a type of ring. Twentieth century mathematicians redefined fields as "commutative rings in which we can divide by any nonzero element."[33]

While studying at the University of Göttingen, Bernhard Riemann (1826–66) attended a seminar on experimental physics destined to influence his research in mathematics. In laboratory experiments, he discerned a close connection between the phenomena of electricity and magnetism. He became convinced that he could develop a single mathematical theory to describe both of these forces.[34] He concluded that both forces were attributable to the bending of space.

Riemann's theory that space can be curved revolutionized geometry by opening up the possibility that space might contain more than three dimensions. He believed that his novel approach to geometry would benefit physics.[35] If anything, he underestimated the ramifications of his approach. In 1915, Albert Einstein drew on Riemann's ideas to build his theory of general relativity. Einstein used the idea of curved space-time geometry to describe gravity.[36]

Riemann is known for his study of curved surfaces and his matrix-like notations, where his work overlapped into linear algebra and group theory. A major source of links between geometry and algebra grew out of applications of "Riemann surfaces."[37] Dedekind used Riemann's work on surfaces to develop a new algebraic approach to geometry.

William Hamilton was the first to use the term "vectors" in algebra. Vectors, a concept imported from astronomy, physics and geometry, made the journey into algebra in the mid-nineteenth century.

Hamilton's original efforts at multiplying 3-dimensional vectors did not fit within Dedekind's conception of fields because of the commutative law. As noted earlier, by suspending the rule of commutativity, Hamilton stumbled across quaternions, one of the earliest algebras discovered. For this, he has been lauded by generations of algebraists and historians of mathematics.

Arthur Cayley (1821–1895), who was 16 years Hamilton's junior, became a mathematician by chance. While training to become a lawyer, he heard Hamilton lecture on quaternions in Dublin. This was a life-changing event for him, and Cayley wound up as a professor of pure mathematics at Cambridge.

He took Hamilton's four-dimensional vectors and transformed them into "matrices," that is, he arranged Hamilton's "numbers" in rows and columns. He described the trick in his paper, "A Memoir on the Theory of Matrices."[38]

Cayley's device not only simplified the solution of equations but created a new mathematical "object." He found that, using his new object, he could multiply as many as eight dimensions.

Mathematicians rewarded Cayley by naming his eight-dimensional vectors "Cayley algebras."

One of the reasons for the steep learning curve in mathematics is that Cayley was not the only mathematician recognized in this way. The lexicon of mathematics is studded with the names of the mathematical stars whose discoveries advanced the science.

Cayley's invention was no minor feat. His work helped to lay the foundation for quantum mechanics, the branch of physics that explains the behavior of subatomic quantities. Matrix multiplication would provide the mathematical framework for the science when it developed some 65 years hence. E. T. Bell gave Cayley's invention of matrices the highest acclaim, noting that "a trivial notational device may be the germ of a vast theory having innumerable applications." [39]

The work of these mathematicians built the framework for modern algebra. But the actual structure was yet to come.

Chapter 8

Flowering of Abstract Algebra over a Shifting Field

The study of algebra underwent a sea change in the twentieth century. A series of developments exploded the former frontiers of the science, taking it to a level of abstraction and generalization considered light years beyond its prior borders. In the words of one commentator, the power of the twentieth century abstract approach "led us to open up whole new areas of mathematics whose very existence had not even been suspected."[1]

In the first part of the twentieth century, modern algebra began to flower on both the European and American continents. In America, Leonard Eugene Dickson (1874–1954) celebrated the dawning of the new millennium by joining the faculty at the University of Chicago. Three years later, he published a paper defining a "linear associative algebra" over an abstract field. While Hamilton's algebras were shackled to the field of real numbers, Dickson's approach freed algebras from their ties to any particular number field, thereby allowing them to range freely over any field of an algebraist's choosing.

The Scottish-born J. H. M. Wedderburn (1882–1948) attended the University of Chicago as a graduate student during the academic year 1904–05. At Chicago, he came under the influence of Moore and Dickson, and adopted their abstract approach of working over arbitrary fields.[2]

Shortly afterwards, Dickson and Wedderburn published a series of papers on "hypercomplex number systems."[3] In one of these papers,

Wedderburn came up with some novel and ultimately, influential, theorems to explain the structure of finite algebras.[4] Wedderburn's theorems were absorbed into the mainstream of research in algebra.[5]

Wedderburn's hypercomplex systems were abstract systems containing vectors and vector spaces.[6] Today, the unwieldy term "hypercomplex number systems" is rarely used, having been replaced by the simpler "algebras" or, alternatively, "associative algebras."[7]

The twentieth century signaled a new approach to thinking about numbers. Abstract algebraists became more interested in exploring the "rules" for solving equations than the solutions themselves.[8] Adrian once told a group of elementary and secondary school teachers, "It seems to me that there should be some place in elementary education where you try to put across the idea that mathematical research does not necessarily consist of trying to solve equations."[9]

Some of the rules, or "axioms," that twentieth century innovators investigated were the familiar ones of high school algebra—the commutative, associative, and distributive laws. In the course of their explorations, they often considered "what if" you suspended one axiom and left all the others in place,[10] much as William Hamilton had suspended the commutative law of multiplication in order to develop quaternions.[11] Hamilton's discovery of a noncommutative algebraic system set the stage for many of the discoveries in algebra during the twentieth century.

After Hamilton's success with suspending the law of commutativity, modern algebraists began investigating the associative law of multiplication, which stipulated that "$a \times (b \times c) = (a \times b) \times c$". Once they recognized that Hamilton's algebras obeyed the associative law, algebraists began categorizing all algebras obeying the law as "associative algebras." The implication of the new term was that if algebraists decided to suspend the law of associativity, they could designate their discoveries as "nonassociative algebras."

In addition to experimenting with the rules/axioms/postulates for performing arithmetic operations, twentieth century scientists shifted the emphasis of algebra. They focused on applying their axioms to the abstract "objects" of modern algebra. Progressing from the study of individual numbers to the study of sets of numbers and complex "objects" such as matrices and vectors enabled mathematicians to develop theories applicable to a broader range of phenomena. As a result, abstract algebraists could use a single formula or equation to solve a number of practical problems at once.[12] As modern algebra developed, its results became increasingly powerful.

Adrian Albert came in at the frontier of the newly evolving science and pioneered in modern higher algebra on American soil. Along with his teachers, E. H. Moore and L. E. Dickson, he belonged to the American school of abstract algebra. He would eventually become one of the leaders in the theory of algebras, both associative and nonassociative.

The role played by algebraists at the University of Chicago was highly influential in shaping the future course of the science. E. H. Moore's abstract approach to algebra[*] influenced every leading American algebraist between 1892 and 1945. As a result, the well-known historian of mathematics K. H. Parshall has chosen to regard America's top algebraists during that period as members of the "Chicago School of Algebra."[13]

D. Fenster, who studies the history of American mathematics, notes that the Mathematics Department at Chicago, with its high research standards, held a "unique position" among American institutions that made the University of Chicago "a viable option for Leonard Eugene Dickson and other aspiring American mathematicians, including Oswald Veblen, Gilbert Bliss, George D. Birkhoff, and R. L. Moore, who might otherwise have traveled to Germany for their training."[14]

In Parshall's words, Adrian Albert "succeeded Dickson as the leader of the Chicago algebraic dynasty."[15] Fenster's perspective is similar. She characterizes A.A. Albert as an "heir" to Dickson. "They both led seemingly tireless mathematical lives, producing—and exerting a worldwide influence through—exceptionally large numbers of publications and graduate students."[16]

Besides the acknowledged high standards and influence of the Chicago algebraists, there was one factor responsible for the success of the "Chicago School". Developments in Germany led to the need for institutions where academic thought could flourish freely. The twentieth century witnessed a shift from Germany to America as the world center of mathematics.

While Moore, Dickson, and Wedderburn forged ahead with major advances in algebra on the American continent early in the twentieth

---

[*] See discussion of Moore's abstract approach in Chapter 3.

century, other brilliant mathematical minds were innovating on the European continent. The University of Göttingen, where the brilliant David Hilbert attracted the attention of scholars from throughout the world, had become one of the most respected centers of mathematical learning in Europe.

Political unrest in Germany followed the Germans' surrender at the end of World War I. Germany's obligations to pay reparations for the damage done to France and Belgium strained the German economy to the point of chaos. National bankruptcy was imminent.

French and Belgian troops marched into Germany after reparations ceased. Concessions allowing Germany to postpone its international debt were made, but failed to staunch the hemorrhaging of Germany's economy. German workers went on strike.[17] Hilbert's former pupil, Hermann Weyl, refused the offer of a post at Göttingen in 1922 due to "the uncertainties of post-war Germany."[18]

Adolph Hitler was thrown in jail the following year for his part in a conspiracy to topple the state government in Munich. While in jail, he managed to complete a manuscript outlining his nationalist philosophy. He called the book *Mein Kampf* (My Struggle).

As the German economy worsened, Hitler, having served a mere 13 months in jail for treason, built his power base. In the mid-1920s, he formed his personal bodyguard army and the Hitler Youth organization. The Nazi party gathered steam in 1927 at a mass political rally in Nuremburg.

As the decade of the 1920s drew to a close, unemployment in Germany began a startling ascent. Unemployment grew to 8.5% by 1929 and steadily rose to nearly 30% in 1932. As unemployment grew, so, too, did the Nazi party's grip on the German government. In 1928, Nazis controlled 12 seats in the Reichstag. By 1932, the number had mushroomed to 230.

The October 1929 stock market crash on Wall Street affected Germany more severely than any other European country. Hitler, who assumed office as German Chancellor in January 1933, blamed Germany's economic woes on the Jews.

The German university system differed from America's in one important respect. While premier universities in the United States were founded as private institutions, Germany's universities were state-owned and funded. As civil service employees, German academics served at the pleasure of the government.

Hitler's vendetta against Jewish professors began in the springtime of 1933. On April 7, 1933, the regime promulgated the "Law for the Reorganization of the Civil Service" requiring removal of Jews from civil service positions.[19] With the edict, Jews were effectively pronounced enemies of the State. Hitler's propaganda machine whipped up public sentiment against Jewish professors by planting rumors that they were part of a vast, anti-nationalist conspiracy to sabotage the new government.

Einstein saw the handwriting on the wall before the others. From his safe berth in Pasadena, California, where he had been wintering in the expectation of returning to Germany in the spring, he announced in March 1933 that he would not be returning to Germany because "I shall live only in a country where civil liberty, tolerance and equality of all citizens before the law prevail....These conditions do not exist in Germany at the present time."[20] Hitler's minions pounced upon Einstein's announcement as evidence that Jewish scientists were traitors.

The action against Jewish professors progressed insidiously. At first, they were placed on temporary leave with full pay. Within about half a year's time, Hitler's regime had converted the temporary leaves into permanent dismissals.

Jewish mathematicians on the faculty at Göttingen were dismissed along with Jewish scholars at universities throughout Germany. By the fall of 1933, the vital flame of mathematical prowess that symbolized Göttingen barely flickered.

When asked by the new Nazi Minister of Education whether mathematics at Göttingen had suffered after removal of Jews from the faculty, Hilbert reportedly replied, "It hasn't suffered, Herr Minister. It just doesn't exist any more."[21]

The mania that swept over Germany spread into attitudes about mathematics. Top scientists, whether Jewish or not, were told that their style of mathematics sounded "too Jewish." In 1936, Germany premiered a new scientific journal named *Deutsche Mathematik*. The new journal would be purged of any Jewish mathematics.

PART THREE

Years of Discovery

Chapter 9

Explorers in Vector Space

Following up on the discoveries of Dickson and Wedderburn, younger mathematicians set out to unlock the secrets at the heart of algebras. Like the discoveries of Hamilton and Cayley, the abstract work of these trail blazers would become the underpinnings of future technologies.

As a graduate student at Chicago in the late 1920s, Adrian Albert gravitated to algebras much as the youthful Bill Gates would later take to computer software. To Adrian, algebras resembled enormous multidimensional labyrinths, just waiting to be explored.

The labyrinths that so fascinated him were more involved than David Hilbert's nineteenth century rings. Ring theory permitted only two operations—addition and multiplication. Algebras posed greater challenges than ordinary rings because a third operation, scalar multiplication, was added to the mix, linking ring theory to vectors. Scalars gave algebras more power than ordinary rings.

The complex numbers and Hamilton's quaternions were the first algebras studied by 20th century mathematicians. Since vector multiplication in these algebras obeyed the associative law, modern algebraists referred to them as "associative algebras."

Adrian believed the theory of associative algebras to be one of the most important branches of modern algebra.[1] From his point of view, associative algebras arose from the study of square matrices, that is, square arrays of numbers or other elements having exactly as many rows as columns.[2]

Matrices are useful creatures because, while they can contain any quantity of elements, they can be added or multiplied together as if they were a single number. This permits mathematicians to reach powerful results with a minimum number of equations. Although valued by pure mathematicians as intrinsically beautiful, matrices have many applications, for example, in the quantum mechanics of atomic structure[3] and CDMA (code division multiple access) digital telecommunications technology.

Adrian began studying matrices early in his career. He employed them to show how the multiplication table of an algebra can be derived, using a factor system.[4]

Hamilton had learned the hard way that the usual assumptions about algebras failed when solving problems involving more than two dimensions. He struggled for 15 years before figuring out how to multiply in three and four-dimensional space.[5]

Adrian's associative algebras, unlike those of Hamilton and Cayley, were not limited to four or eight dimensions. Abstract algebraists like Adrian experimented with multiplying in "n" dimensions, where "n" could be any positive integer.

Today's physicists are still debating the issue of whether our world includes more than four dimensions (three dimensions of space and one of time). Mathematicians were ahead of the curve—they posited "n" dimensions in their equations without waiting for proof in the physical world.

One of the main questions on the minds of algebraists in the first part of the 20th century was "What are all of the possible algebras?"[6] That was too massive a question to contemplate. Wedderburn became the first to formulate an approach to simplifying the problem.

Early in the 20th century, he proved that the problem of determining all of the associative algebras could be reduced to the classification of "division algebras," that is, algebras having the property that division (except by zero) is always possible.[7] Wedderburn considered division algebras to be "the ultimate building blocks" of associative algebras.[8]

What Wedderburn's main theorems did was to show that if you only knew the division algebras over a given field, you could essentially describe *all* of the algebras over that field.[9] He noted that one might as well consider division algebras as "algebras over their centers" and so assume that they are "central" in the sense that their center is the "base field."[10] Hence, the term "central division algebras" came into use.*

Wedderburn proved, while still in his twenties, that finite division algebras were fields. His result later found applications in other branches of mathematics such as number theory and projective geometry.[11]

Until 1905, the only known division algebras were fields and Hamilton's quaternions over the real numbers.[12] No new division algebras had appeared on the horizon since Hamilton's sudden intuition in 1843. Dickson broke the logjam in 1905 when he discovered a brand new class of division algebras. This amazing new class appeared to be endowed with universal properties.

Dickson's epiphany went largely unnoticed at the time. There were no reports of champagne parties to celebrate the event or of cars overturned on the Midway in the course of reveling. Wedderburn, one of the few scholars to recognize the significance of Dickson's discovery, dubbed the newborn creatures "Dickson algebras." It was only with the benefit of hindsight that Adrian was able to proclaim 30 years hence, "Such algebras are probably the most important types of algebras."[13]

Wedderburn showed in 1921 that Dickson algebras were special cases of a more general type of algebra now considered a "crossed product."[14] Since Dickson algebras were crossed products defined by "cyclic fields," algebraists found it more useful to refer to Dickson algebras as "cyclic algebras."

In 1923, Dickson compiled all of the results that he and Wedderburn had developed on the theory of algebras and presented a unified approach in his treatise *Algebras and Their Arithmetics*. Dickson's seminal tome delineated the properties of "central division algebras" and "algebraic numbers," and showcased his Dickson (cyclic) algebras. He laid down an implicit challenge to future scholars by listing three major unsolved problems in algebra: the determination of all division algebras, the theory of nonassociative algebras, and the extension of algebras to the whole theory of algebraic numbers.†

---

* "Central" division algebras were referred to as "normal" division algebras at that time.

† According to Dickson, "when the coefficients of an algebraic equation are all rational numbers, the roots are called algebraic numbers."

Dickson decried the fact that the evolution of algebra had been "surprisingly slow." Filled with the moxie of youth, Dickson's disciple was in a hurry to put algebra on a fast track. Beginning in 1928, Adrian set his sights on Dickson's first challenge—determining all division algebras, which Adrian called "the chief outstanding problem" in the theory of associative algebras.[15]

Abstract algebraists of that era viewed division algebras in much the same fashion as today's geneticists view the human genome. In order to "determine" the division algebras, one needed to investigate their structure. Algebraists hoped to eventually determine that all division algebras shared a particular structure, and to discover its nature.

After achieving an understanding of the overall structure of division algebras, the next objective would be to break division algebras down into separate classes. By easing up on a given condition, a mathematician might discover a new class. Eventually, they hoped to discover all possible classes of division algebras.[16]

Adrian's early papers expressed his frustration with the mathematical community's "meager knowledge" of the structure of division algebras.[17] He wondered whether, in determining all central division algebras, he might discover that they were all Dickson-type cyclic algebras. Proving their cyclicity would yield a concrete and explicit understanding of the structure of division algebras even to the point of revealing their multiplication table.

Dickson's research had proved that all 4-dimensional (2^2) central division algebras are cyclic, while Wedderburn proved that the same characteristic holds true for 9-dimensional (3^2) central division algebras. Adrian approached the next case scenario in his second doctoral thesis topic, where he explored the properties of 16-dimensional (4^2) central division algebras. In the thesis, he proved that 16-dimensional central division algebras are not necessarily cyclic algebras, but are always crossed products.[18]

He fine-tuned his results and presented them at the April 1928 meeting of the American Mathematical Society in Chicago. The following year, the AMS published a revised version of Adrian's dissertation. This was the 22-year old's first major contribution to science.[19]

Kaplansky observed, "Here was a tough problem that had defeated his predecessors; he attacked it with tenacity till it yielded. One can imagine how delighted Dickson must have been."[20]

Another mathematician intimately familiar with Adrian's work, Daniel Zelinsky (a former Albert doctoral student), put the doctoral dissertation in perspective when he concluded: "Although Albert's theorem raised the obvious question about algebras of dimension 25, 36, etc., his result has stood without essential improvement or embellishment (though not for lack of trying) until some nice, complementary, but still not definitive results...in 1971."[21]

In his search to determine all of the division algebras, Adrian studied the work of the German mathematician Ferdinand Georg Frobenius (1849–1917). Frobenius may have been the first to consider the case of division algebras over the real numbers. Frobenius theorized in 1878 that the only division algebras over the field of real numbers were: the field of real numbers themselves, the field of complex numbers, and the algebra of quaternions.[22] Dickson gave a proof of Frobenius' theorem in his treatise *Algebras and Their Arithmetics*.[23]

Adrian Albert's explorations would not be tethered to a specific number field, such as the real number field investigated by Frobenius. He set out to explore division algebras over the field of rational numbers under the theory that, if one could discover these algebras, the result could be generalized to apply to *any* field of numbers. His search for answers lurking in the field of rational numbers proved to be right at the heart of the developing structure theory of algebras between 1929 and 1932.

The rational number field presented far greater challenges than the field of real numbers considered by Frobenius. It would take more perseverance to unravel its secrets.

In 1929, mathematicians suspected that all division algebras over the rational number field met Dickson's test for cyclicity, since no one had found any counterexamples. Adrian attempted to test that assumption. He constructed a 16-dimensional central division algebra over the field of rational numbers in a paper presented to the AMS in December 1929.[24] Front and center, he posed the question of whether any non-cyclic central division algebras exist. After a thorough examination, he concluded that, to date, "no non-cyclic algebras are known to exist," but then proceeded to set forth some criteria to be met for algebras to qualify as non-cyclic.[25] A reader skimming the paper might easily miss the important conclusion regarding non-cyclic algebras, since it appeared in the middle of the paper.

The enigmatic paper marked the beginning of an international controversy that continues among mathematicians to the present day. The question concerning Dickson's cyclic algebras gradually evolved into a now-famous mathematical theorem. As Hitler's march to power gathered steam, a fervent search to develop a proof began simultaneously in Germany and America. The controversy went beyond the mere issue of who found the proof. There were innuendos of possible duplicity on the part of a German mathematician arising out of his desire to secure glory for his native country by claiming credit for the discovery to the exclusion of his American counterpart.

Perhaps the controversy would not have generated so much heat had it not been for the events that followed. The plot seemed to thicken after the German mathematician became an important figure during the Third Reich, while the American went on to develop encryption methods used in World War II. When I learned of the controversy, I decided to get to the bottom of it by tracing the steps of the participants as the story unfolded.

Chapter 10

The Proof

L.E. Dickson updated and expanded his treatise, *Algebras and Their Arithmetics*, in 1927 for a German language edition titled *Algebren und ihre Zahlentheorie*.[1] The publication overseas of Dickson's book created a buzz among German mathematicians about the revolutionary ideas of the pioneering Chicago professor.[2] German mathematicians were particularly interested in Dickson's notions about the relationship between the theory of algebras and the theory of algebraic numbers.[*] The marriage of these seemingly incongruous partners held the fertile promise of many mathematical progeny. Talk about Dickson's book soon spread across Germany's mathematical community, and sparked a flurry of activity overseas.

One of the first German mathematicians to appreciate the advances wrought by Dickson and Wedderburn was David Hilbert's protégé and colleague, Fraulein Amalie ("Emmy") Noether. In 1919, she began working on ideal theory at the University of Göttingen. Noether analyzed the work done by Dedekind, Hilbert, and Wedderburn relative to rings and attempted to unify and simplify the subject. In 1921, at the age of 39, Noether made her mark by publishing a paper titled *Idealtheorie in Ringbereichen* that established the framework for modern ring theory.

---

[*] In the introduction to his 1923 book, *Algebras and Their Arithmetics*, Dickson summarized his approach as follows: "Adopting a new definition, the author develops at length a far-reaching general theory whose richness and simplicity mark it as the proper generalization of the theory of algebraic numbers to the arithmetic of any rational algebra."

By the time Dickson's book exploded onto the mathematical scene in Germany, Noether was well versed in the developing abstract theory of algebras. She began advocating in favor of applying the theory to other branches of mathematics.[3] In a 1925 lecture, Noether demonstrated that Wedderburn's structure theorems provided a useful framework for developing the "representation theory of groups" that was becoming a hot topic among German mathematicians.

In 1927, Noether was in close communication with both Richard Brauer, a former doctoral student of Frobenius at the University of Berlin, and Helmut Hasse, a former student of hers and Hilbert's who wrote his doctoral thesis under Kurt Hensel. She managed to persuade Brauer, who was investigating "representation theory," that the study of algebras could be used to advance his research.[†]

Over a period of years, Noether also persuaded Hasse, whose research focused on algebraic number theory, that "her methods," that is, applying the theory of algebras, would benefit his research. Hasse became increasingly interested in the theory of algebras once he recognized the connection between this developing theory and his own studies in "class field theory," a branch of algebraic number theory.[4]

Just as Adrian Albert was placing the finishing touches on his enigmatic paper about the cyclicity of algebras over the rational number field, half a world away, Hasse began to focus on Dickson's invention. Hasse came up with some conjectures about the cyclicity of algebras over number fields and wrote to Noether in December 1930, confessing to her that his conjectures were not yet supported by a solid foundation.[5]

Noether replied to Hasse, "Yes, it is a terrible pity that all your beautiful conjectures are floating in the air and are not solidly fixed on the ground: for part of them—how many I do not yet see—hopelessly crash through counterexamples in a very new American paper...by Albert."[6]

[†] Noether's advice had a lasting impact on Brauer's research. Describing Brauer's approach to group theory during the 1960s, the mathematician Daniel Gorenstein observed that "Brauer had come to finite group theory via algebraic number theory and the theory of algebras." (D. Gorenstein, "The Classification of the Finite Simple Groups, a Personal Journey: The Early Years," in *A Century of Mathematics in America*, vol. 1, at p. 454.)

She did not yet accept the notion that every division algebra whose center is an algebraic number field was necessarily a cyclic algebra.[7] In retrospect, this fact is surprising, since it shows that even a brilliant innovator such as Emmy Noether considered the novel idea to be pie in the sky.

After receiving her letter, Hasse read the Albert paper and decided that she had misconstrued it. Nonetheless, to resolve any remaining doubts about whether the young American mathematician had disproven his conjectures, Hasse decided to contact the source of his uncertainty.

An intercontinental dialogue began in January 1931 with a letter from Hasse to Dickson expressing an interest in the work of Dickson's protégé. Adrian was flattered by the attention from the older, well-established mathematician, and sent this cordial reply from his office at Columbia University on February 6:

> I was very interested to receive your letter from Professor Dickson describing your work. I am very pleased at your interest in my work and am sending you a set of my reprints.... I have also five new papers being published and shall send you the reprints as soon as I obtain them.

Both Hasse and Albert stood to gain from sharing their work. The library at Columbia was deficient when it came to certain German mathematical journals. Adrian had heard about the work of the great Emil Artin, one of the developers of "class field theory." Correspondence with Hasse offered broad access to the ongoing work of top German mathematicians. At the same time, Hasse wanted to learn the latest findings of Dickson's former student, who was blazing new trails in his investigations into the structure of algebras.

Hasse got right to the point. In his first direct communication with the young American, Hasse followed up on the puzzling paper. Hasse wanted to know for certain whether Adrian Albert had in fact discovered any non-cyclic 16-dimensional division algebras.[8] He replied that he had investigated the question in the paper "of which you have a copy," but, "The question seems to be a number-theoretic one and I see no way to get an algebraic hold on it."[9]

Hasse undoubtedly felt validated. His conjectures had not crashed after all, but remained standing and ready to forge ahead. Afterwards, the correspondence between the two men blossomed into an intricate *pas de deux* as they shared further developments in their research.[10]

The youthful American worked alone on his side of the Atlantic, constructing theorems and proofs, coming ever closer to deciphering the puzzle of cyclic algebras over number fields. Although he worked independently, Adrian kept Dickson and Wedderburn apprised of his progress.

On the other side of the Atlantic, the 33-year old Hasse was also hard at work on the problem. Unlike his American counterpart, Hasse did not work alone. Two of the most brilliant mathematical minds on the European continent, Noether and Brauer, continued to share their results with him, as they had done since 1927.

None of the participants enjoyed the luxury of proximity to one another. Noether corresponded from her post at Göttingen with Hasse, who had accepted a professorship at Marburg University beginning with the summer of 1930. Brauer occupied a post at the University of Königsberg. Collaboration among the trio was facilitated by the efficiency of the German postal system—mail between these German cities generally reached its destination within a day or two.

Adrian produced a series of papers between 1929 and 1931 that solved various aspects of the problem.[11] While the American launched a piecemeal attack on the problem in his papers, the German trio did the same. Mathematical brainwaves on both sides of the Atlantic were operating on the same frequency.

Although fraught with delay, Adrian's letters to Hasse were the most efficient means of keeping the German mathematicians apprised of his results. There was no trans-Atlantic Airmail, let alone a worldwide web to deliver instantaneous messages. In the 1930s, letters from America took approximately 2 weeks to cross the Atlantic. Sending newly published papers to Hasse posed still greater obstacles. Many months often elapsed between Adrian's submission of an article for publication, the referee process, typesetting, and his receipt of reprints.

As the duo exchanged their results, each took some care to arrange for publication of his respective findings as the research progressed. Both knew full well the importance of achieving priority, since a mathematical journal would not publish a discovery once it was published elsewhere. At the same time, on several occasions, Adrian shared with Hasse unpublished research in the making. For example, Adrian responded to a query from Hasse about "direct products" and "matrix algebras" by sharing some solutions in two back-to-back letters written in March 1931.[12]

Subsequent observers have felt compelled to speculate about the mindsets of the participants. To the German mathematician Peter Roquette, a former student of Hasse, the German team-leader and the American were engaged in a collaboration involving teamwork and free exchange of results. In contrast, Kaplansky saw the relationship as a competitive race to the finish line.[13] We will search for a resolution to the debate as we follow the story.

Adrian developed a powerful new tool for simplifying the problem in a paper he submitted for publication on April 17, 1931. He succeeded in reducing the problem to the case involving prime power degree, using a theorem he called the "index reduction factor theorem."[14]

After completing the paper, but before publishing it, he learned of similar results developed by Brauer that were virtually unknown in America. Adrian duly credited Brauer in his paper while demonstrating that his own independent proof was more streamlined and contained some new and different results. He mentioned the coincidence in a letter to Hasse, noting:

> ...my own proof uses none of the theory of "factor systems," is self-contained, and is rather short. Brauer's work, on the other hand, depends on two previous memoirs and two papers of I. Schur. Wedderburn thinks my method is a great advance and so I am publishing my new proof.[15]

In hopes of avoiding another instance of overlapping results, Adrian made this request:

> The work of the German mathematicians on algebras is very interesting to me and I should like to know all of it if possible. In particular, I have been unable to obtain Artin's Hamburg papers. Can you tell me in a few words what subjects he studied in these papers on linear algebras, and whether or not he has published anything in more accessible journals? I certainly do not wish to repeat known results, even if they are unknown in America, and

I am very pleased and thankful for the opportunity to communicate with you and know of your results.[16]

In the spring of 1931, Adrian succeeded in advancing the research on rational division algebras by proving that all 16-dimensional central division algebras over the field of rational numbers are in fact cyclic (Dickson) algebras.[17] He detailed his proof, along with some other new findings, in a May 11 letter to Hasse.[18]

Hasse wanted to use the new proof in a paper that he planned to submit to the American Mathematical Society's *Transactions*, and wrote Adrian for permission to do so.[19] Adrian, who had already succeeded in arranging for publication of his proof in the June 1931 *Proceedings of the National Academy of Sciences*[20] ahead of submitting a more detailed paper for the *Transactions*, replied, "I should be pleased to have you refer to my result in your Transactions paper."[21]

The summer of '31 was a busy one for both men. The prior year, England's Royal Society had asked Hasse to comment on the credentials of Wedderburn as a candidate for membership in the Society. Hasse gave his comments in German because he felt his English skills inadequate to pay proper tribute to Wedderburn. Now, Hasse had added motivations for improving his English—comprehending the work of American mathematicians and publishing in American scientific journals.

Hasse decided the best approach to polishing his English was to invite an Englishman to stay at his home for a few months. At Hasse's request, an English professor sent a young mathematics student, Harold Davenport, to live with the Hasse family during the 1931 summer semester on condition that Davenport speak only English to Hasse.‡ By the end of the summer, the Hasses and Davenport had become such good friends that they embarked on a road trip to tour Europe together in Davenport's car. We can infer that Hasse had informed Adrian of his summertime plans, as he replied to Hasse on June 30:

> I hope to hear from you again, and I shall also write to you at the end of this summer when I may perhaps have more things of interest to you. I have some new results

---

‡ The author thanks Professor Peter Roquette of the Mathematics Institute at the University of Heidelberg for this anecdote.

now but wish to wait until they are more complete before making any communication of them to you. P.S. Your English is very clear and understandable. I only wish I could write German half so well!

Adrian ended his postscript with a wistful thought about the future: "I hope that in perhaps two years I may visit Germany and there see you and discuss our beautiful subject, linear algebras."[22] In the course of their correspondence, a spirit of collegiality had developed between the two men who shared the bond of a love for pure mathematics.

While Hasse was busy with his houseguest, Adrian Albert labored on the proof. Adrian's knowledge of the algebraic mathematical techniques of Dickson/Wedderburn combined with his recent exposure to the arithmetic methods of Hasse and Artin caused the solution to gel in his mind over the summer months. By September, Adrian had achieved a major breakthrough. He could sense the final pieces of the puzzle falling into place. On September 9, 1931, he transmitted several new algebraic theorems to the AMS for publication in the October issue of the *Bulletin*.[23]

As Hasse would ultimately acknowledge, A.A. Albert's new theorems led directly to solution of their "principal theorem" that "every finite dimensional central division algebra over an algebraic number field‖ is a cyclic (Dickson) algebra."[24] Jacobson described Albert's achievement as follows: "Albert obtained independently all of the algebraic results on splitting fields, extensions of isomorphisms and tensor products which were needed to obtain the fundamental theorems on division algebras over fields."[25]

Hasse finally found the time to write to Adrian in mid-October. The German mathematician remained in the dark about the new theorems in the American's latest paper, since Adrian had as yet received no reprints to send him. Nor could Hasse access the paper in Marburg, since American mathematical journals appeared later in Europe than in America.

While the precise contents of Hasse's October letter are unknown, he was still several steps away from completing the proof.[26] However,

---

‖ "An algebraic number field is a field whose quantities are ordinary complex number roots of equations with rational coefficients." (A.A. Albert, *Structure of Algebras*, at p. 129.)

from Hasse's description of his latest findings, Adrian discerned that Hasse, too, had made dramatic inroads on the solution.

As he turned Hasse's letter over in his mind, Adrian began to conceptualize the remaining theorems needed to complete the proof. On November 4, he wrote to Hasse congratulating him on achieving "such an important result" and continued, "...your result just communicated gives me almost immediately the following theorems."[27] Whereupon Adrian proceeded to share three new, unpublished, theorems that he derived, building on Hasse's latest result.

According to the subsequent account of Hasse and Albert, the theorems in Adrian's November 4 letter came very close to completing the proof of their principal theorem.[28] Adrian concluded his letter with this question:

> When and where do you intend to publish your new result? Of course, my own theorem is not proved unless I can refer to your paper? Can we not perhaps make some arrangement to publish in the same place (or even in the same paper!)?[29]

The clock was ticking down to the final solution. When he wrote to Hasse in early November, Hasse's proof remained incomplete. It took the combination of letters sent to Hasse by Brauer and Noether on October 29 and November 8, respectively, to give Hasse the ideas he needed for the ultimate solution.[30] It finally dawned on Hasse how their ideas could be used to simplify the problem, and he completed the proof on November 9, 1931.

Here is where the story takes a sharp turn. Neither the theorems published by Adrian in September nor the ones in his November 4 letter had made their way across the Atlantic before Hasse took his next step. Within a day's time, Hasse hurriedly slapped together all of the pieces of the proof to produce a paper titled "Beweis eines Hauptsatzes in der Theorie der Algebren" ("Proof of a Main Theorem in the Theory of Algebras"). On November 11, Hasse submitted the paper to himself as editor of a leading German mathematical journal, *Journal für Mathematik*.[31] The paper gave the names of only two coauthors—Brauer and Noether. In newspaper vernacular, Adrian had been scooped.

The very day he submitted the paper, Hasse wrote to Adrian about his forthcoming publication. He apparently felt the need to justify his reasons for failing to include the American as one of the theorem's co-developers. Hasse attempted to show that the results in Adrian's April 17 paper were almost all "not new." He went on to point out other aspects of the proof, which he ascribed to German mathematicians.

Two weeks later, Adrian received Hasse's letter and immediately recognized the impact that the Hasse-Noether-Brauer paper would have on the mathematical community once it was released. Posterity would not know of his own parallel solution. Since his major contribution to the proof was a milestone for the twenty-something prodigy, he was determined to claim credit for work done on his side of the pond.

He dispatched a lengthy letter to Hasse (dated November 26), congratulating him on completing the proof while rebutting some of Hasse's assumptions. Recognizing that Hasse had not seen his latest work, Adrian provides Hasse with the text of the new breakthrough theorems appearing in the *AMS Bulletin*'s October issue. He explains the similarity between his new theorems and a portion of the proof that Hasse's paper attributed to Brauer and Noether, concluding, "As my theorems have already been printed I believe that I may perhaps deserve some priority on that part of your proof...."[32]

In response to Hasse's rationale that the results in Adrian's April paper regarding the "index reduction factor theorem" were "not new," Adrian counters with a detailed explanation of the innovations that Hasse missed in the paper, and observes, "Note also that my paper was received by the editors of the *Transactions* on April 17, 1931, yours, presented April 24." Finally, Adrian demonstrates how some of the results that Hasse ascribed to German mathematicians were in fact traceable to earlier papers by himself and Dickson.

While claiming priority for some of his own work, Adrian was filled with praise for the work of the German team. With regard to the proof developed by Emmy Noether only days after his own proof, he offers:

> It is of course impossible for people working in the same field not to frequently obtain overlapping results. I have the highest respect for Fr. Noether and certainly appreciate her great mathematical powers. Please carry to her my most sincere best wishes and appreciation for her accomplishments in our beloved field of ALGEBRA.[33]

Concerning Hasse's achievement in obtaining the so-called "existence theorem" relating to cyclic fields, Adrian refers to it as "remarkable."[34]

He closes by conveying regards from Professor Dickson, thanking Hasse for communicating the "wonderful theorem," and adding this handwritten postscript:

> I am very glad that you are interested in the possibility of my visiting you. I hope that I will be able to leave Chicago on September 1, 1933 to return here not later than December 31, 1933. I do not believe I can make the trip before that time.

Although he named Noether and Brauer as his only coauthors for the *Journal für Mathematik* paper, Hasse had dropped in a footnote crediting Adrian for developing the "index reduction factor theorem." After reviewing the substance of Adrian's latest work, Hasse expanded the footnote in galley proofs by adding the following:

> ...A. A. Albert's results show that he has an independent share in the proof of the main theorem. Moreover, Albert has remarked (after having seen our proof of the main theorem) that our central theorem 1 follows in a few lines from theorems 13, 10, 9 in a paper of his which is currently printed (Bull. Amer. Math. Soc. 37 (1931)).[35]

In Hasse's defense, it must be said he was obliged to take it on faith that Adrian's new theorems had already appeared in print, given the fact that reprints were not yet available. However well-meaning Hasse's amended footnote may have been, it was small solace to the man who knew that American mathematicians would learn about the German trio's triumphant success while remaining unaware of the obscure footnote in the German-language journal.

On December 9, Adrian wrote to Hasse after reviewing proof sheets of Hasse's paper, noting, "I feel even more strongly now that in E. Noether's and R. Brauer's part of your paper there was a good deal of unnecessary complication due to the fact that the...reduction was not made."[36] Adrian believed his own reduction to be simpler than the reduction developed by Noether and Brauer.

Hasse responded by offering to set the record straight by means of a joint paper laying out their respective contributions to the proof, and presenting a new, more concise, proof developed by Adrian.[37] At Hasse's suggestion, Adrian drafted the article. Having completed it within a month's time, he mailed a draft to Hasse with a letter indicating that he planned to send it on to the editors of the *Transactions* soon.[38]

Hasse approved Adrian's draft subject to several revisions made by Hasse in proof sheets prior to publication.[39] In their joint article, they conclude that "proof of the main theorem is an immediate consequence of Hasse's arithmetic and Albert's algebraic results."[40]

Once the joint Albert-Hasse paper appeared, some in the mathematical community, having gained a new understanding of events, began referring to the discovery as the "Albert-Brauer-Hasse-Noether theorem."[41] Others persisted in calling it the "Brauer-Hasse-Noether theorem," despite Hasse's subsequent acknowledgment of Adrian's independent share in the proof.

This brings us back to the original controversy concerning mathematical priority, mindset, and morality. The controversy involves Hasse and Albert alone, since they were the pivotal actors in the drama.

With regard to priority, there is general agreement that, although Hasse was first to publish the complete proof, Adrian's independent results overlapped substantially with the essential breakthroughs by the German team. For example, Roquette drew the following conclusions:

> It is not an unusual mathematical story. A major result is lurking behind the scenes, ready to be proved, and more than one mathematician succeeds....We conclude that Albert's Bulletin paper and the Brauer-Hasse-Noether paper had been written independently of each other....as Albert points out correctly, his results in his Bulletin paper can be used to prove the Local-Global Principle for algebras, and hence indeed constitute an "independent share" in the proof of the Main Theorem.[42]

As to whether the German and American scholars viewed themselves as competitors or collaborators, the truth lies somewhere in between. On the one hand, both men were pleased to share their ideas and results in a mutually beneficial manner. Adrian's November 4th proposal that they coauthor a joint paper shows that he considered the project to be a collaborative effort. By then, he had been corresponding with Hasse over a period of 9 months, sharing not only his papers, but also his unpublished ideas, as he went along.

Was the civil servant, Helmut Hasse, influenced by the mounting force of German nationalism? Piecing together bits of history, Roquette provides a rationale for Hasse's rush to publish—Hasse's desire to meet the deadline for a special issue of *Journal für Mathematik* honoring the 70th birthday of his mentor, Kurt Hensel.[43] In other words, Hasse was anxious to include the discovery in one of Germany's premier scientific journals featuring an issue dedicated to a German mathematical icon. Would listing an American mathematician as coauthor have detracted from the acclaim surrounding Hasse's announcement in the German journal? For his part, Adrian proceeded with his own research, publishing his significant findings from time to time, demonstrating a healthy regard for his own priority. Both men appear to have teetered on the fine line between competition and collaboration.

There remains, in the ongoing debate about the controversy, the innuendo that Hasse might have behaved in an underhanded manner towards Albert. Did Hasse deliberately deprive Albert of credit for his role in developing the proof? Did the older, more seasoned, Hasse take undue advantage by seizing an opportunity to publish first?

The evidence tends to show that he did not. The factor of a substantial delay in communicating Adrian's results meant that Hasse had not received the theorems from the October *Bulletin* paper or the November 4 letter in time to make use of them. Moreover, it was Hasse's suggestion that they coauthor an English-language report of their respective discoveries in order to allay any remaining doubts about who had discovered what.

Subsequent correspondence from Adrian settles the question of whether he harbored any ill feelings resulting from the chain of events. In a January 1932 letter, Adrian provides Hasse with some of his latest research and closes with this handwritten sentiment:

Permit me to say I did not believe it possible for mere correspondence to arouse such deep feelings of friendship and comradeship as I now feel for you. I hope that you reciprocate.[44]

As Adrian had anticipated, resolving the structural issue as to the cyclicity of algebras yielded the key to understanding associative algebras in general. Proving the theorem opened the way to a complete classification of division algebras over number fields.

In addition, the solution came with a significant fringe benefit. Roquette notes that it opened "new vistas in one of the most exciting areas of algebraic number theory at the time, namely, the understanding of class field theory."[45]

When Emil Artin heard that Hasse and the others had completed their proof, Artin became so excited that he wrote to Hasse, "You cannot imagine how ever so pleased I was about the proof, finally successful, for the cyclic systems. This is the greatest advance in Number Theory of the last years...."[46]

Looking back on the struggle to unravel the structure of algebras, A.A. Albert later reflected, "The theory of linear associative algebras probably reached its zenith when the solution was found for the problem of determining all rational division algebras."[47] Much of his subsequent treatise, *Structure of Algebras*, flowed from that essential discovery.

Meanwhile, he proceeded to build on the newfound revelation. The results he and the others obtained on the structure of algebras suggested a related problem: is every finite dimensional central division algebra of prime degree cyclic?[48] This problem concerned algebras over fields of characteristic p where p is a prime number and where the dimension of the algebra over its field is a power of this same prime p.[49] The Albert-Brauer-Hasse-Noether breakthrough had not settled this question because number fields are of characteristic zero, not of prime characteristic.[50]

He spent a great deal of effort on this problem in a series of papers. In Jacobson's view, Albert's papers on the theory of "p-algebras" were a "beautiful chapter in the structure theory of central simple algebras."[51]

Once again, here was a topic of mutual interest to Albert and Hasse, and they continued their intercontinental dialogue via snail mail. The intensity of Adrian's passion for his craft shines through his letters. He declares: "It really is a burning question with me as to whether the theorem is true." And later: "I believe the above theorems are very beautiful and hope they are worthy of work by other mathematicians...."

In a paper published in January 1932, he reduced all problems on cyclic algebras to the case where the exponent n is a power of a prime p.[52] He believed his reduction to be simpler and more elegant than previous techniques, and wrote to Hasse, "I cannot understand why you still insist on working with cyclic algebras of degree n instead of degree p^e, a prime...of course this is a matter of personal taste and you may not agree."[53] Hasse penned in the margin of Adrian's letter, "for example class field theory is as simple for n now!"

Adrian then chided Hasse for his blanket conclusion that every division algebra similar to a cyclic division algebra is cyclic.[54] Hasse scribbled a marginal note, "yes! error of me!"

After many years of effort, Adrian managed to prove that "every p-algebra is cyclically representable, that is, similar to a cyclic algebra."[55] However, he reserved judgment on whether any non-cyclic p-algebras exist and whether the fact that p-algebras are cyclically representable implies that they are indeed cyclic. This question remains open seven decades later.[56]

The question that Dickson's book sparked in 1923 continued to mystify mathematicians. Now that the principal theorem was resolved, did it mean that all central division algebras were cyclic? Adrian continued to pursue the issue, and became the first to resolve it. In the spring of 1932, he finally cracked the problem by proving that non-cyclic 16-dimensional algebras do in fact exist.[57] He regarded this as one of his signal achievements.[#]

[#] See "Albert, Abraham Adrian," *McGraw-Hill Modern Men of Science* (1966), at p. 6. Although there followed some controversy regarding whether Richard Brauer had found the solution first, Brauer himself in a 1933 paper acknowledged Albert's priority. (R. Brauer, "Uber den Index und den Exponenten von Divisionalgebren," *Tôhoku Mathematical Journal* 37 (1933), 77-87.) Roquette concurred in the conclusion that "The first who has explicitly constructed a non-cyclic division algebra was Albert."

Unfortunately, all but two of Hasse's letters to his American colleague have disappeared, and one can only infer the contents of the remainder from Adrian's replies. What emerges is that the correspondence between the two men displayed a genuine mutual respect and admiration.

Neither Hitler's rise to power on January 30, 1933, nor the German government's edict placing Jewish university professors on leave in the spring of 1933, affected Adrian's wish to continue corresponding with his German colleague. Likewise, his exposure to the flood of German refugees he met at the Institute for Advanced Study during the fall of 1933 failed to stifle his correspondence with Hasse. In a December 1933 letter, he offered to coauthor a new paper with Hasse if Hasse could complete one of his proofs. The American sent at least four letters to Hasse in the 12-month period between February 1933 and February 1934.

Afterwards, the equation shifted. A.A. Albert, who rarely allowed external events to interfere with his pursuit of mathematical research, could no longer avoid noticing that the world was changing.

Richard Courant, a German Jew, had fought valiantly in the German army during World War I and sustained serious battle wounds. After leaving military service, he produced quality mathematics that drew the attention of the German mathematical community. In 1920, he received an appointment to the faculty in Göttingen where he had earned his doctorate.

Courant later succeeded Felix Klein as head of Göttingen's mathematics faculty and fulfilled a longstanding dream of Klein's—making Göttingen's first mathematics building a reality. Klein had laid the plans and arranged the financing, but it was Courant who brought those plans to fruition. He lovingly shepherded construction of Göttingen's glorious new Mathematical Institute, which opened in December 1929. However, he was ousted from his position as director within weeks of the Nazi regime's announcement of the "Law for the Reorganization of the Civil Service." He spent the next year teaching in Cambridge, England and then fled to New York City with his family in January 1934.[58]

During the summer of 1934, Hasse assumed the directorship of the Mathematical Institute at Göttingen following the rapid dismissals of a series of directors who were either of "non-Aryan" descent or failed to demonstrate that they were sufficiently pro-Nazi to satisfy Hitler's regime. Oxford University's Marcus du Sautoy describes the aftermath of a turbulent period in Göttingen's history as follows: "During Hitler's destruction of the mathematics department in Göttingen, Hasse was appointed by the Nazis to take over the running of the department. Hasse's Nazi sympathies, combined with his mathematical abilities, made him a suitable candidate in the eyes of the authorities and the mathematicians in Germany, who hoped to preserve Göttingen's tradition."[59]

Hasse's friend Davenport tried to persuade him that the Nazi regime's policy of firing all Jewish professors was immoral. Hasse responded by rationalizing Hitler's expulsion of Jewish professors as a temporary "sacrifice" needed to restore the German state.[60] He longed for Germany's return to the glory days of 1871[61] when the German empire was established following a series of military conquests, and confided to Davenport that he had in fact voted for the Nazis.[62] Davenport eventually ceased all efforts to persuade him, and their strong differences of opinion drove them apart.

This leads us to consider what became of the friendship between Albert and Hasse. Hasse continued to express an interest in Adrian's research, and penned a 21-page letter to his American counterpart early in 1935. He wrote to inquire about a 1934 paper of Adrian's containing a conjecture that could potentially streamline the "existence theorem" used in their famous proof.[63] Hasse added, "I have read your recent papers on Riemann matrices with *greatest interest*, also Weyl's new paper on this. We are studying the matter in my Seminar."[64] No subsequent correspondence between the two men has been found.

While it does not appear that Adrian replied directly to Hasse's letter, he took steps to sustain a somewhat tenuous relationship. Carl Ludwig Siegel, an Aryan mathematician who spent the 1934–35 academic year at the Institute for Advanced Study, informed Hasse after returning from Princeton, "A. A. Albert has asked me to convey to you his regards."[65] Hasse thanked Siegel for the regards and mentioned his interest in Albert's results on Riemann matrices.[66] We do know that Adrian received the Hasse letter, since he referred to it in a 1938 paper where he disproved the conjecture that Hasse inquired about.[67] The vibrant two-way communication between the two men had degraded into sporadic communication through third parties and oblique references in printed journal articles.

Having returned to Germany in the fall of 1935, Siegel was devastated to learn that some of his closest colleagues at the university in Frankfort had been forced from their posts. In 1940, Siegel left Germany because he could no longer tolerate the policies of Hitler's regime. He would remain in self-imposed exile for the next eleven years.

Hasse, on the other hand, grew more militant in the late 1930s.[68] Since he owned up to a Jewish branch in his family and had ostensibly warm relationships with his Jewish doctoral advisor Kurt Hensel and with the Jewish collaborators in his mathematical papers, Hasse's transformation is not well-known.

Dissension erupted after the German publishing house, Springer, introduced the Nazi regime's agenda into the editorial policies of an international mathematical journal, the *Zentralblatt für Mathematik*, thereby unalterably compromising its impartiality and scientific integrity. Springer chose to eliminate as editors and referees not only Jewish mathematicians, but Aryans who had emigrated from Germany after Hitler came to power. American mathematicians began debating among themselves whether to establish a competing journal in America and the effect such an action might have on relations with their German friends.

Hasse, whose words carried special weight due to his position as head of Germany's premier Mathematical Institute, wrote to the prominent American mathematician Marshall Stone about the matter in March 1939:

> Looking at the situation from a practical point of view, one must admit that there is a state of war between the Germans and the Jews. Given this, it seems to be absolutely reasonable and highly sensible that attempt was made within the domain of the *Zentralblatt* to separate the members of the two opposite sides in this war. I do not understand why the American mathematicians found it necessary thereon to withdraw their collaboration in bulk. I do not know whether it was the intention, but it certainly has the appearance of taking decidedly and emphatically one of the two sides, and thus deviating from a truly impartial and hence genuinely international course....
>
> It is no good crying over spilt milk. One had better wipe it away and avoid spilling more.[69]

He concludes with the warning: "The foundation of an equivalent to the *Zentralblatt* in America would, under the present circumstances, severely aggravate a situation already grave enough."

After more than a month's reflection, Stone sent this statesmanlike reply:

> Your remarks concerning the *Zentralblatt* impressed me as so important and so authoritative that I felt obliged to transmit those of a general and impersonal nature to the Committee which is considering the American position in the matter. I trust that you will not regard my action as inconsiderate or incorrect in relation to yourself.
>
> From your letter, I infer that some important elements in the American position are not appreciated in the German mathematical community. It appears to me that a decision to inaugurate a new abstracting journal in the United States is not likely to be taken for the purpose of passing judgment on the past history of the *Zentralblatt*. If taken, it will be based primarily on the desire to assure for the future of mathematical abstracting a responsibility commensurate with America's great and growing mathematical importance. In view of the possibility— whether remote or imminent I shall not undertake to say— of a destructive upheaval in Europe, the American mathematicians are wisely considering whether America might be a safer haven to an international journal than most other lands....
>
> ...Springer was under no legal obligation to consult the interests of the American group or any other. His unqualified demands upon the editor both disclosed and decided the issue in so abrupt a fashion as to preclude conciliatory efforts on the part of the collaborators, however impartial. To my mind there is no occasion for surprise that many collaborators felt themselves forced to take stands *pro* or *contra* on so sharply drawn an issue. As for the Americans who withdrew, I feel that they merely acted with true loyalty to their own national traditions and ideals.... Personally I regard the matter as closed.... Rather than cry over spilt milk, I prefer to look

for a reliable container which will spill as little as possible in the future.[70]

Hasse, who had volunteered for naval service during World War I, returned to naval duty from 1939 until 1945, working in Berlin on ballistics.[71]

Chapter 11

Traveling Riemann's Universe

Bernhard Riemann rejected the old flat-world Euclidian notions about a universe of parallel lines that never intersect. Instead, he embraced a new paradigm where space is curved like the surface of a sphere.[1]

Riemann developed a theory about the relationship of geometry to the real world. One commentator illustrates Riemann's conception of the universe by describing the hypothetical behavior of bookworms placed on a crumpled sheet of paper. The bookworms, seeing only the part of the paper in their immediate vicinity, would get the impression that the paper was flat, and would not notice the distortion. However, when they attempted to move across the crumpled paper, they would feel a mysterious force preventing them from moving in a straight line. "Riemann concluded that electricity, magnetism, and gravity are caused by the crumpling of our three-dimensional universe in the unseen fourth dimension. Thus a 'force' has no independent life of its own; it is only the apparent effect caused by the distortion of geometry."[2]

The methods Riemann invented to explore forces at work in space enabled future mathematicians and physicists to overcome the constraints of geometry. He introduced new approaches to the theory of functions. During the 1850s, he developed a matrix approach as a device for applying function theory to the investigation of physical phenomena. Riemann's matrices corresponded to his conception of a world comprised of curved surfaces.

Adrian first learned of some long-unsolved problems involving Riemann matrices in that idyllic setting at Princeton, where the 22-year old and his bride drank in the exhilaration of a new venture. To the young couple away from home together for the first time, life seemed full of possibilities.

Within weeks of their arrival in 1928, the postdoctoral research fellow delivered his first lecture at the Princeton Mathematics Club. Professor Solomon Lefschetz sat in the audience, his finely tuned mental gears churning. Listening to the young man speak about his doctoral dissertation on division algebras, Lefschetz grasped the connection between Adrian's work and a well-known problem in Riemann matrices.* After the lecture, Lefschetz proceeded to describe the problem to him as the two men wandered the streets of Princeton together.

Adrian might never have come across the problem on his own, since it appeared to be outside his area of concentration. Lefschetz himself had wrestled with it from the vantage point of an expert in algebraic geometry. Building on the work of the great French mathematician Henri Poincaré and others, Lefschetz made several significant advances in the study of Riemann matrices and Riemann surfaces.

Lefschetz was one of the first to discern the potential of a purely algebraic approach to the analytic and geometric problems raised by Riemann's matrices.[3] In 1919, he won recognition for providing the most complete classification theory of Riemann matrices and their multiplier algebras up to that time.

The early research done by Lefschetz provided the springboard for Adrian's work in this area. In the words of a leading geometer, Lefschetz' work "predestined the nature of [A.A.] Albert's results and in doing so exerted an important influence on the theory of associative algebra."[4]

Soon after Lefschetz introduced him to the topic, Adrian proceeded to scrutinize some of the puzzles that had remained unsolved for more than half a century.[5] He focused on a pair of objectives—determining the structure of all pure Riemann matrices and gaining a definitive understanding of their multiplication algebras.

Like a juggler keeping multiple balls aloft, he produced several papers on the subject between 1929–31 while working at solving the cyclicity theorem for associative division algebras.[6] As he progressed in his investigations into the structure of algebras, Adrian increasingly

---

* See discussion regarding Adrian's introduction to Riemann matrices in Chapter 4.

observed links between that topic and the problems of Riemann matrices.

He began to make substantial progress on Riemann matrices only after the cyclicity problem was solved. His work on the structure of division algebras turned out to be a critical ingredient in cooking up a solution for this new puzzle.

Zelinsky described the task as follows: "What was required was a classification of the algebraic correspondences of a Riemann surface (automorphisms of a complex curve). This had been reduced to the problem of finding the matrices that commute with a certain 'Riemann matrix' of periods of basic Abelian integrals on the Riemann surface. These commuting matrices form an algebra, and in the basic cases, a central simple algebra over the rational field. This version of the problem was right in the center of Albert's expertise, and he demolished it."[7]

Jacobson explained how Adrian went about producing a solution: "Albert's work on Riemann matrices went hand in hand with the development of the theory of division algebras. It culminated in the complete solution of the principal problem, which he published in three papers appearing in the *Annals of Mathematics* in 1934 and 1935….To achieve this required the development *ab initio* of the basic theory of simple algebras with involution."[8]

"Involution" involves mapping of an algebra to itself, that is, associating each element x of an algebra with an element x' in the same algebra.[9] In Adrian's hands, involution became a tool for problem-solving.[10] He is credited with "the classification of Riemannian matrices via algebras with involution, and in fact virtually the entire theory of simple algebras with involution…."[11] His application of involution theory to simple algebras provided the architectural framework needed to complete the structure theory of Riemann matrices.

In 1934, Adrian finally determined the precise structure of the multiplication algebra of a pure Riemann matrix. He followed up that feat by discovering the existence of a Riemann matrix consisting of algebraic numbers.

This time around, there was no controversy over who had found the solution. After six years of work and approximately a dozen papers on the subject, he earned the undiluted praise of the mathematical community. For his work on the long-unsolved problems in Riemann matrix theory, the American Mathematical Society awarded him the prestigious Frank Nelson Cole Prize in Algebra. The thirty-something scholar finally had the recognition he craved. There was no disputing his mathematical prowess.

The rare prize was awarded once before—a decade earlier, Adrian's mentor, L. E. Dickson, received the first Cole Prize in Algebra. The prize was to be awarded in five-year intervals, but the awards committee could not agree on a worthy recipient in 1934, so the award to A.A. Albert in 1939 was only the second ever awarded. The next two recipients were Zariski in 1944 and Brauer in 1949.

The prize was founded in honor of a Harvard mathematics professor best remembered for the mathemagical performance he gave at a meeting of the AMS in 1903. Standing before the audience of AMS members, Cole scribbled some numbers on one blackboard, stepped to the right, scribbled some more numbers on another blackboard, placed an equals sign between them, and sat down. The equation he had just written was:

$$2^{67} - 1 = 761838257287 \times 193707721.$$

The room immediately erupted in thunderous applause as the entire assemblage rose to its feet. Cole's equation provided the solution to an age-old problem in number theory involving the prime numbers. It was known that $2^{67} - 1$ was not itself a prime number, but the product of two smaller prime numbers. However, mathematicians did not know which ones. Being the first to factor $2^{67} - 1$ was a noteworthy achievement in number theory.[12]

Given Cole's solution of an age-old mathematical problem, it made perfect sense that the prize partly endowed by Cole should be given to mathematicians who succeed in solving famous or difficult problems.

After Adrian's use of involutions in studying Riemann matrices, involutions evolved into a separate branch of the theory of algebras.[13] *The Book of Involutions*, a massive tome on the subject, points out it was A.A. Albert who "laid the foundations of the theory of simple algebras with involutions."[14]

Noting that "[t]he first systematic investigations of central simple algebras are due to Albert," *The Book of Involutions* draws heavily on his early work.[15] The text is peppered with references to "Albert forms," "Albert quadratic space," and "Albert algebras." The authors conclude that, in working out the theory of Riemann matrices, he completely determined the structure of "multiplication algebras."[16]

Until the 1930s, Adrian led a rather sheltered life, untouched by the ravages of war and government-sponsored vendettas. It came as a shock to him when he learned that the Nazi regime had introduced a political agenda into his beautiful and pure subject, algebra. At the same time, he appreciated the irony behind the belief held by Hitler's minions that there was such a thing as "Aryan mathematics."

Oswald Teichmüller, a doctoral student of Helmut Hasse, actively promoted this misguided notion. At Göttingen, Teichmüller gained a reputation as a scientifically gifted man who was "completely muddled and notoriously crazy."[17] A ringleader of pro-Nazi students, he succeeded in the ouster of a brilliant professor named Edmund Landau on the ground that Aryan students did not wish to hear any lectures about Jewish mathematics.

Teichmüller was among the young mathematicians who published their research in Germany's new all-Aryan journal, *Deutsche Mathematik*.[18] Adrian read one of Teichmüller's *Deutsche Mathematik* articles with a mixture of anger and amusement. Teichmüller's paper on cyclic algebras contained a valuable new proof. However, the youthful firebrand had not created his proof in a vacuum. It hinged on a proof that Adrian had published four years earlier, and Teichmüller used Adrian's proof in the article.[19] Teichmüller wrote the paper while enrolled in Hasse's seminar, and most likely learned about the Albert proof from Hasse.[20]

The hypocrisy of portraying the journal as strictly "Aryan mathematics" insulted Adrian's sense of fairness. He later showed the article to his doctoral student, Daniel Zelinsky. With a wry smile, he pointed to the journal. "This Aryan journal contains a Jewish proof!"†

A.A. Albert recognized that abstract algebra was so new, it was incomprehensible to both established and beginning mathematicians. In his first textbook, *Modern Higher Algebra*, published in 1937, he sought to give students a foundation for understanding recent breakthroughs in associative algebras.[21] Lacking such a foundation,

---

† Sanford Segal, in his tome, *Mathematicians under the Nazis*, testifies that A.A. Albert was one of ten Jewish mathematicians whose works were cited in *Deutsche Mathematik*.

students could not fully grasp these new discoveries, much less, build upon them. It was his way of opening up the field of mathematics to the next generation of mathematicians.[22]

Through his contacts in the American Mathematical Society, the American Council on Education learned of the Chicago professor's efforts to improve the level of mathematics education in American universities. In 1933, the Council appointed him to a committee on graduate instruction seeking to identify schools offering adequate Ph.D. programs in mathematics and rank those with the best programs.

The American Mathematical Society had helped to launch his mathematical career by recognizing the value of his doctoral dissertation and publishing the young scholar's maiden effort in its then premier journal, *Transactions of the AMS*. He viewed the Society as a venerable organization, and remained active in it throughout his life.

He was elected to sit on various AMS Editorial Committees beginning in the mid-1930s. In 1934, the AMS selected him as a Symposium lecturer. The following year, he was elected to membership in the AMS Council, the Society's governing body.[23]

The success of the Colloquium held following the mathematical congress associated with the 1893 World's Fair led to the now much-celebrated "Colloquium Lectures" delivered by an outstanding mathematician at annual meetings of the Society. Following presentation of the addresses, the AMS often publishes them in hard cover volumes.

After his first book appeared in 1937, the AMS designated Adrian "Colloquium lecturer" for an upcoming Colloquium.[24] He was determined to uphold the high standards of the historic tradition. E. H. Moore had delivered the Colloquium Lectures in 1906, and Dickson presented them in 1913.

Preparing his lectures gave Adrian the impetus to finish the job he began in his first textbook—explaining the revolutionary changes that were transforming modern algebra. He struggled to prepare them ahead of the 1939 Colloquium so that participants would have his new text in hand at the meetings.

He arrived to deliver the lectures in a state of anxiety, since the books he hoped to distribute were nowhere to be found. The printers finished their work with scarcely enough time to ship them directly to the conference site at the University of Wisconsin. He confided to Frieda,

I did not write yesterday because I was too nervous. The book wasn't here when I arrived....Then they did get it and I was excited....The book is fine in appearance if a bit thin. I have been congratulated repeatedly on getting it out on time.[25]

Attendees thought it a great feat that he managed to get the book into print ahead of the lectures, since no one had ever done that before.[26] He titled the collection *Structure of Algebras*.[27] Coming some 16 years after Dickson's book on the general theory of algebras, Adrian's text can be seen as a sequel. It superimposed a structure on the foundation built by Dickson and Wedderburn by incorporating developments that were virtually bursting the prior confines of knowledge in abstract algebra. *Structure of Algebras* received kudos from mathematicians and earned a reputation as a classic treatise on the theory of linear algebras.[28]

Another feather in his cap came when A.A. Albert was appointed in 1939 as one of four editors of the *AMS Bulletin*, a journal featuring brief papers about new developments in mathematics. Each of the editors was responsible for the papers in his particular field.

The job of Managing Editor fell to Dean Tomlinson Fort of Lehigh University in Pennsylvania. Although the position accorded him a certain status in the mathematical world, it came with significant headaches. Budgetary constraints required the Managing Editor to crack the whip on his colleagues to keep the length of papers to an absolute minimum. To avoid budget overruns, paper quality had to be cheapened twice, and a backlog of papers awaiting publication was becoming a serious problem.

On at least one occasion, Fort proposed to J.R. Kline, Secretary of the Society, that two authors who asked to have their papers withdrawn from the *Bulletin* after type had already been set be charged for printing costs.[29] Kline replied that authors assume a responsibility when submitting their papers and that, although the matter lay in the Managing Editor's discretion, charging authors under those circumstances would be a wise policy.[30] Fort followed up by instructing the central office staff to bill the authors and notify him if they had any trouble collecting the debt. He then notified the governing Council of the AMS, "This procedure has the approval of the editors of the *Transactions* as well as of the *Bulletin*."

While Adrian and the other editors did most of their editorial work from their home offices, the job entailed occasional trips to the east coast. After completing his duties at AMS offices in New York, he often stopped off in Princeton to visit old acquaintances. In October 1941, he penned the following letter to his beloved Frieda and mailed it in an envelope bearing the return address "Bulletin of the American Mathematical Society, A. A. Albert, Editor":

> Dearest: (not to mention my own sweet darling)
> I finally solved the problem at 6 A.M. It's right this time. I'll write it up a bit later....I had lunch with the Bochners yesterday and dinner with the Lefschetzs...I'm having a grand time and feel well. I told Veblen no soap about the Institute. So that's over. Both he and Lefschetz think I'm right not to go at the present time....I saw Wedderburn and he seems much better....xxxx to all of you. Your Adrian.

After serving four years on the Editorial Board of the *Bulletin*, he learned that his name was being bandied about for the position of Managing Editor of the *Transactions*, the AMS' premier scientific journal. Unlike the *Bulletin*, the *Transactions* featured full-length research papers.

He confided to Kline, "I am exceedingly eager for the Transactions Editorship."[31] Garnering the position would be a milestone in his career. At the time, he viewed the appointment as an honor and gave no thought to Fort's travails or the potential grief accompanying the job of a Managing Editor.

Soon after the appointment as head of the *Transactions*, his efficiency as a manager became evident. He repeatedly cut through layers of red tape and outmoded rules to get mathematical papers to press as quickly as possible. As a result, delays in publication were reduced substantially.[32]

During his first year as *Transactions* editor, he received the electrifying news that he should stop the presses and prepare to publish a paper by a German-born mathematician named Hans Rademacher. Rademacher claimed to have solved a famous problem considered the holy grail of prime number theory—the Riemann Hypothesis.[33]

The hypothesis was invented by the same genius who developed Riemann matrices. Riemann studied numbers like 2, 3, 5, 7, 11, 13, and 17, which cannot be expressed as the product of two smaller integers.

Mathematicians had observed that prime numbers distributed themselves in a random, unpredictable fashion. Riemann came up with a hypothesis that attempted to deliver order to the seeming chaos of the sporadic prime numbers. He conjectured that the frequency of prime numbers is closely related to the behavior of an elaborate, but predictable, function.[34] Mathematicians have been attempting to prove or disprove Riemann's conjecture ever since he constructed it in the mid-19th century.

Rademacher's paper claimed to have disproved the hypothesis. The discovery was considered so significant that it made the pages of *Time* magazine. However, C. L. Siegel, who refereed the work, found a fatal flaw in the paper that invalidated it. Afterwards, Adrian expressed the disappointment of the mathematical world when he noted, "The whole thing certainly raised a lot of false hopes."[35] The behavior of prime numbers has remained as elusive to mathematicians as the behavior of the mysterious cicadas who disappear beneath the earth and regularly reappear at 13 or 17-year prime number intervals.

Adrian stepped down as *Transactions* editor in 1949, and, for much of the 1950s, served as editor of the AMS hardcover series, *Colloquium Publications*. In later years, he continued his service by holding elective offices in the Society that played a pivotal role in American mathematics.

Next to Chicago, Adrian's closest affiliations were in Princeton. Oswald Veblen, one of E. H. Moore's former doctoral students, was instrumental in organizing the Institute for Advanced Study and became a professor there. Veblen attempted to lure Adrian back to the Institute for the 1940–41 academic year.

By that time, his life had become more complicated. He now had three children instead of one and a household full of possessions. He wrote to Veblen declining the offer, explaining that he had pondered "long over a decision which, several years ago, would have been made promptly in your favor."[36] What he omitted from the letter was that, in the interim, his four-year old son Roy had contracted juvenile diabetes following a bout with measles and the disease was very difficult to manage.

The University of Chicago rewarded him for remaining on campus that year by giving him only advanced classes to teach and four months entirely free of teaching duties. Still, he felt conflicted about turning down a chance to spend another year at the Institute, and was delighted when the Institute's John von Neumann wrote inviting him to spend a week in March 1941 lecturing as a guest of the Institute.[37] He offered to speak on something "fresh and new," the topics of "extensions of p-adic fields" and "non-associative algebras."[38] In the ensuing years, he would return to the Institute periodically to lecture on new developments in his research.

The Institute continued to entice Adrian to return for another year. Over a period of years, he went back and forth with Oswald Veblen, Hermann Weyl, and other Institute faculty about the possibility of returning to spend another year there, but it was not to be. The advent of World War II, the effect of the War on the Institute's finances, the Alberts' trips abroad, and the defense work that took him to other locations all stood in the way.

Plans were approved for him to spend the summer of 1955 at the Institute conducting basic research in linear algebras under a grant from the Office of Naval Research. The Institute's Marston Morse wrote to Dr. Robert Oppenheimer, "He is a leading algebraist whom it would be good for us to have around."[39] The plan would have allowed Adrian time for pure research after a heavy load of administrative duties in connection with chairing the National Research Council's mathematics division. However, he eventually opted out of the plan in favor of working on a summer research project in Southern California.

Since he had gained a reputation for his work on associative algebras early in his career, mathematicians often mentioned algebra and Albert in the same breath. Chicago's mathematics students were quick to spoof the connection. The Mathematics Department customarily held an annual Christmas party where students staged a skit, playing the roles of faculty members and giving them comical names. One skit featured an A. Adrian Albert impersonator with the monicker "A. Associative Algebra."[40]

Chapter 12

A New School of Algebra

It was not the first time that Adrian Albert picked up the seed of an idea in Princeton and planted it firmly in midwestern soil. He reaped a bountiful harvest after learning about Riemann matrices from Lefschetz during his postdoctoral year.

His return visit to Princeton in the fall of 1933 did not disappoint, although that would not become apparent until years later. Who knew that his attentiveness at Hermann Weyl's lectures would affect not only Adrian's life, but the lives of his mathematical children and grandchildren?

Weyl was 48 years old when he arrived at the Institute for Advanced Study during its opening year as a permanent member. He had learned his lessons well from Hilbert, who taught Weyl that a student absorbed nothing before hearing it five times.[1] The shy Weyl became an inspiring teacher.

Weyl's talents were wide-ranging. He made contributions to topology, group theory, quantum mechanics, and number theory, and produced a significant theory regarding the geometrical properties of space-time. In the course of his investigations, Weyl developed an interest in the work of the Norwegian geometer, Sophus Lie (1842–1899).

Lie's study of geometry led to his work in establishing a correspondence between lines and spheres by a process called "contact transformations" which allowed tangent spheres to correspond to intersecting lines. Lie was greatly influenced by his interactions with leading algebraists of his day and found their techniques to be useful. To establish the correspondence, Lie employed "partial differential equations." In the process of combining contact transformations, Lie discovered an amazing algebra so unique that mathematicians named it after him. "Lie algebras" have since found applications in various branches of physics, such as supergravity and quantum mechanics.

During Adrian's 1933–34 stay at the Institute, Weyl delivered a series of lectures on "Lie groups" and "Lie algebras" that introduced Lie's ideas to an American audience. Weyl, noted for his astute philosophical observations about mathematicians and mathematics in general, once commented: "The individual mathematician feels free to define his notions and set up his axioms as he pleases. But the question is, will he get his fellow mathematician interested in the constructs of his imagination?"[2]

By implication, Adrian answered the question. Weyl's lectures on Lie algebras captured the young American mathematician's imagination. As Jacobson later noted, the lectures "aroused Albert's interest in this subject—an interest which later broadened to encompass the whole range of nonassociative algebras..."[3] (those algebras where the law of associativity is not assumed).

In that nutrient-rich cauldron of ideas, the Institute for Advanced Study, Adrian was exposed to a second type of "nonassociative algebra." The German physicist Pascual Jordan was searching for a new algebraic framework for quantum mechanics.

The mathematical framework used relied on a cumbersome type of matrix multiplication.[4] Jordan was seeking an algebra that would avoid the pitfalls of the existing approach.

Jordan developed an algebra over the field of real numbers (an "r-number algebra") that he hoped would lead to a solution. He then proceeded to identify the axioms that applied to his new "r-number algebra."

To aid in his search for a complete algebraic solution, Jordan enlisted the help of the mathematician John von Neumann and the mathematical physicist Eugene Wigner. Their goal was to classify all categories of Jordan's "r-number algebra."

In a joint paper, the trio broke Jordan's algebra down into four infinite classes and one exception.[5] The "exception" was a strange number system that appeared to be the algebra they were seeking, but they could not prove it. They asked the young temporary member at the Institute, Adrian Albert, to look into it.[6]

Adrian's expertise in the structure of algebras proved to be the key to deciphering the enigma cloistered within the realm of physics. After mulling the matter over in his mind, he hit on a theorem to explain the strange object. As a result, he invented a "new algebra" in a paper that he submitted in 1933.

As Jordan and the others had conjectured, the new algebra proved to be an "exceptional" creature quite different from any algebra previously encountered. It refused to obey the axioms that Jordan had identified for the four "r-number algebras."[7] Princeton University's journal, *Annals of Mathematics*, published Adrian's article back-to-back with the paper by Jordan, von Neumann, and Wigner in January 1934.

At first, Jordan and the others were pleased that he had discovered the "exceptional algebra" they were seeking. However, their hopes were soon dashed when they concluded that Adrian Albert's algebra was too tiny to provide an adequate home for quantum mechanics.[8]

Afterwards, his "exceptional algebra of quantum mechanics" lay dormant for almost two decades. At the time, it appeared to be a useless curiosity of no interest to the mathematical community.

After his first venture into nonassociative algebras in 1933, A.A. Albert returned to the subject on the eve of America's entry into World War II. In two papers published together in 1942, he laid the foundation for a new school of mathematical thought by exploring the "general theory" of all nonassociative algebras.[9]

Kaplansky viewed the events leading up to publishing the first paper to be as significant as the paper itself. "The date of receipt was January 5, 1942, but he had already presented it to the American Mathematical Society on September 5, 1941, and he had lectured on the subject at Princeton and Harvard during March of 1941. It seems fair to name one of these presentations the birth date of the American school of non-associative algebras, which he single-handedly founded."[10]

Adrian aptly described the state of the art in 1941 when he explained, "The study of non-associative algebras has already yielded much of interest and importance. Indeed those *special theories* [the so-called 'alternative algebras,' Lie algebras, and Jordan algebras] in which the associative law is replaced by a substitute theory have each been of an extent and of an interest almost comparable to that of the theory of associative algebras." He went on to observe, "The results on non-associative algebras in which one does not assume a type of partial associativity...have almost...all been of a rather primitive kind and have been scattered through the literature."[11]

What made Lie algebras, Jordan algebras, and alternative algebras so "special" was that they resided in a never-never land suspended between associative and nonassociative algebras. Those "special theories" obeyed axioms that served as surrogates for the associative law; they assumed a "partial associativity." His idea was to reach beyond those much-touted "safe" topics by exploring algebras in which no postulates of associativity were assumed. This was uncharted territory, and Adrian had no way of predicting whether his explorations would bear fruit.

His investigation would hurdle the wall between the worlds of associative and nonassociative algebras to explore what was on the other side. He wanted to uncover the essential structure of *any* nonassociative algebra. It was as if he said, "Give me any nonassociative algebra. I will tell you how to take it apart and tell you what it really looks like—I will reveal its fundamental structure to you."[12] The ultimate goal was to discover as much about nonassociative algebras as mathematicians already knew about associative ones by undertaking a comprehensive study of this unexplored area.

He investigated virtually every facet of nonassociative algebras. For starters, he introduced the concept of "isotopy," an equivalence relation imported into algebra from the study of topology.[13] The concept involves one-to-one mappings of sets with a unity quantity. Isotopy later proved to be useful in "non-Desarguesian geometry."[14]

Adrian wrote several papers on "general nonassociative theory." He also studied a number of classes of nonassociative algebras defined by their identities.[15] He defined the "radical," that is, "root," of any "finite dimensional nonassociative algebra" and demonstrated its divergence from the radical of an associative algebra.[16]

In 1952, he invented a nonassociative division algebra he called a "twisted field."[17] Later on, he generalized the concept[18] and defined an entire class of finite division algebras he called "twisted fields." He employed the new class of algebras in his studies on "combinatorial analysis," "finite projective planes," and "non-Desarguesian projective planes."[19]

Apart from his study of the "general theory" of nonassociative algebras, Adrian explored the various "special cases" of nonassociative algebras. During the mid-forties, he decided to explore the "r-number algebra" that Jordan and the others had investigated in connection with quantum mechanics.

Rather than consider the problem from the perspective of physics, he took up the subject from a purely algebraic point of view. In a paper written in 1945, he used the term "Jordan algebras"* to refer to this special case of nonassociative algebras.[20] Then, he launched an effort to unravel the inner workings of Jordan algebras with an intensity similar to his previous efforts on behalf of associative algebras.

Kaplansky told the story of Adrian's breakthrough discovery in Jordan algebras as follows:

> I arrived in Chicago in early October 1945. Perhaps on my very first day, perhaps a few days later, I was in Albert's office discussing some routine matter. His student Daniel Zelinsky entered. A torrent of words poured out, as Albert told him how he had just cracked the theory of special Jordan algebras. His enthusiasm was delightful and contagious. I got into the act and we had a spirited discussion. It resulted in arousing in me an enduring interest in Jordan algebras.[21]

Adrian followed up by publishing three papers on the subject, including a paper providing a structure theory for Jordan algebras.[22] According to Kaplansky, "These three papers created a whole subject; it was an achievement comparable to his study of Riemann matrices."[23]

He took the road less traveled when it came to mathematics research. J. Marshall Osborn, a former Albert doctoral student, described his teacher's approach to selecting research topics: "Albert was unusual among algebraists in that he was willing to look at almost any system. He didn't have sharp limits. Albert was one of two stars. Jacobson was the other star, but he was interested only in Jordan algebras, the best-behaved nonassociative algebras. Jacobson was not interested in anything weaker than that."[24]

He took off on a tangent by investigating "quasigroups" in two papers during World War II.[25] These objects were not considered algebras because they involved groups rather than rings.

---

* Adrian had actually coined the term "Jordan algebras" in a footnote to his 1942 paper on fundamental concepts of nonassociative algebras.

He became intrigued with a quasigroup having an identity element, and coined the term "loop" to describe it. One could imagine the quasigroup as a loop because the identity element has an inverse, so that it loops around. The Midwesterner confided to some of his students that he picked the term "loop" because he wanted to name it after Chicago.[26] Chicagoans generally associate "Loop" with the city's downtown area, which is rimmed by a loop of elevated train tracks. The name with a Chicago flavor stuck, and mathematicians adopted the catchy phrase, "loops and quasigroups."

Adrian delved into the topic of "power-associative rings" beginning with a paper he produced in 1947.[27] "Associative rings" are rings that obey the associative law. Power-associative rings are rings where "subrings" generated by a single element also obey the associative law.

In a subsequent paper, he showed that "power-associative rings" were related to Lie algebras and Jordan algebras.[28] The Albert paper introduced the notion of "Lie-admissible algebras." This result later found application in research on clean fuel technology.[29]

Afterwards, he developed a body of work on "power-associative algebras," that is, algebras where the "subalgebra" generated by any one element is associative. He presented theorems and a structure theory in a series of papers on the topic.

The Russian mathematician A. T. Gainov later wrote, "His works in algebra represent a valuable contribution to world mathematics and possess numerous continuers. The works of Professor A. A. Albert on the structural theories of power-associative algebras were of a special interest to me."[30]

Adrian next turned his attention to "alternative algebras," that is, those algebras where the subalgebra generated by any two elements obeys the associative law. In 1948, he explored "right alternative algebras."[31] The following year, he invented a new type of algebra that he called an "almost alternative algebra."[32]

One of his most significant contributions to alternative algebras was his classification of "simple alternative rings" with "idempotents."[33] Idempotents are rather peculiar elements having the property "$e^2 = e$."† In other words, the element multiplied by itself equals itself, for

---

† Benjamin Peirce first defined the term "idempotent" in 1870. Dickson introduced this simplified definition in his 1923 text, *Algebras and Their Arithmetics*.

example, "$1^2 = 1$". Making vital use of papers that Adrian published in 1949 and 1952, Erwin Kleinfeld broke new ground by classifying all simple alternative rings in 1953. Upon reading Adrian's papers, Kleinfeld exclaimed, "It's amazing! He proved exactly the right things."[34]

Of the special cases of nonassociative algebras, Lie algebras have found the most practical applications.[35] Adrian had a hand in discovering several new classes of Lie algebras.[36] In 1954, he and one of his doctoral students, Marguerite Straus Frank, produced results on a new class of the "simple" Lie algebras that form the basic building blocks for Lie algebras.‡ [37]

During 1954–55, he worked with Kaplansky on a research project in nonassociative algebra sponsored by the U.S. Army's Office of Ordnance Research. The project also supported three graduate students, including another Albert doctoral student, Richard Block. Block wrote his 1956 doctoral thesis on new simple Lie algebras of prime characteristic. In the 1980s, Block gained recognition for further advancing the subject with some "marvelous work."[38]

A.A. Albert was the first to attempt to organize all possible nonassociative algebras.[39] He summarized his work in nonassociative algebras as follows: "...[I] produced the general structure theory of Jordan algebras, and later gave a construction of exceptional Jordan division algebras....[I] also constructed the major classes of finite nonassociative algebras. This work provided most of the known classes of finite nondesarguesian projective division ring planes."[40] He introduced a generation of young scholars to this newly developing area, which fascinated him.

The originality of the topics he selected for investigation convey the impression that he gave no thought to where his explorations would lead. However, as Kaplansky has pointed out, Adrian "repeatedly displayed an uncanny knack for selecting projects which later turned out to be well conceived...."[41]

Although the two were thought to be unrelated, he managed to bridge the gap between associative and nonassociative algebras by using one to prove theorems in the other. According to Zelinsky, "Albert had the idea of using associative algebra theory to prove

---

‡   Chapter 28 tells the story of their collaboration.

analogs of the Wedderburn theorems for quite arbitrary nonassociative algebras (even a nonassociative algebra has an associative 'regular representation' algebra). It is a sign of his genius that he was actually able to develop a reasonable theory and also significantly influence the theory and applications of the special algebras [Lie, Jordan, and alternative algebras]...."[42]

Commentators have noted Adrian's unique "style" of mathematics. Zelinsky describes it as "inimitable. He had a diabolical facility with manipulation of identities—an enterprise in which most mathematicians founder, never being able to see the forest for the trees. Somehow, Albert could see through mazes of symbols to the inner workings of all those polynomials in several variables or multiplication tables of complicated algebras."[43]

His unusual style became evident during a presentation on some of his findings in Jordan algebras at a weekly meeting of the University of Chicago Mathematics Club. He scribbled a number of equations on the blackboard and came out with some surprising results. Some in the audience wondered what underlying theories had guided his thinking.

Professor Saunders Mac Lane jumped up and demanded to know, "Why *do* the results turn out that way?" Adrian stood silent, wearing a sphinxlike smile. Mac Lane, furious, pounded on the table. "Adrian knows, but he won't tell us!" The truth was, he could not explain how he arrived at his results—they sprang from an intuition he could not put into words.[44]

Like his lectures, Adrian's written works were sometimes inscrutable, even to other mathematicians. His 1956 book *Fundamental Concepts of Higher Algebra* updated and expanded his earlier text, *Modern Higher Algebra*. A reviewer noted that the new book was more than a simple repetition of the old material:

> The selection and arrangement are expertly done, and new proofs are produced for a number of theorems to improve the unity and logical structure of the presentation.

The reviewer tempered his enthusiasm with this observation:

> Unfortunately, the virtues of the book are likely to be appreciated only by the specialist. Albert's style at its softest makes few concessions to the reader.... It is too bad

that the difficulties of style will limit the number of readers who might otherwise appreciate the brilliant qualities of this book. Mathematics could use more writers like G. H. Hardy.[45]

In a retrospective, Jacobson characterized A.A. Albert's mathematical style as "highly individualistic.... He had a fantastic insight into what might be accomplished by intricate and subtle calculations of a highly original character."[46] Jacobson concluded, "Perhaps the most striking of Albert's qualities as a mathematician was his power, especially his fantastic computational skill, and his tenacity in pursuing difficult problems of recognized interest."[47]

Despite the frenetic pace of his scholarly work, he was not an "all-work-no-play" mathematician. He loved fishing in the north woods. He enjoyed "roughing it;" Frieda did not. Twice, he tried bringing the entire family along and renting a cabin for the summer. The second time, I was four years old and nearly drowned.

After that summer, he limited his fishing trips to a week or two at a time, going either alone or with one of his two sons. To him, nothing tasted better than freshly caught fish that he had cleaned, filleted, and sautéed himself. Frieda was glad to stay home with her running water and washing machine. In later years, he worked out an arrangement with the owner of some cottages on Plum Lake in northern Wisconsin to leave his outboard motor on site so that he could revisit the same spot year after year.

He thrived on competition and competitive games. Besides being an avid fisherman, he was a tennis player, golfer, and an enthusiastic billiards, pinochle, and bridge player. In off moments, he played a mean game of cribbage.

The card-playing habits of Mathematics Department members were notorious around the University. In the 1938 production of the faculty club's satire, "The Revels," he portrayed "Mr. A." in a skit entitled "It's Your Lead—An Afternoon of Duplicate."

PART FOUR

Branching

Out

in

Midlife

Chapter 13

Mathematics during World War II

During the 1930s, war clouds were bearing down upon the mathematical community. In February 1933 Adrian informed Hasse, "My trip to Germany is now indefinitely postponed as next year I have a leave of absence here and go to the Institute of Advanced Study in Princeton, on a one year appointment. I don't have any plans for the period after that."[1] Later that year, Richard Brauer and Emmy Noether fled Germany following the Nazi government's edict dismissing all Jews from their posts at German universities.

Brauer and Noether were not the only top Jewish scientists to escape Hitler's wrath. The great exodus of mathematicians had already begun. Albert Einstein's winter respite in the United States had turned into a permanent exile.

Emil Artin fled in 1937 out of concern for his wife, who was Jewish.[2] Although Hermann Weyl's wife was only part Jewish, his friends in America persuaded him to get out of Germany before it was too late.[3] Before leaving, Weyl was recognized as the premier mathematics professor at Göttingen.[4] He became famous for his application of group theory to quantum mechanics.

Among the Jewish scientists who poured out of Europe was the brilliant Hungarian mathematician, John von Neumann, who had trouble finding a good job in Europe due to his Jewish roots. Once in America, he wrote a text providing a mathematical framework for quantum mechanics.[5] Von Neumann's book took a rather crude theory devised by physicists and gave it an elegant mathematical structure. Even today, physicists refer to von Neumann's book as their "quantum bible."[6]

Two Jewish physicists, Leo Szilard and Edward Teller, who had previously escaped an anti-Semitic regime in their native Hungary by immigrating to Germany, managed a quick exit from Germany soon after Hitler came to power.[7] The Hungarian mathematical physicist, Eugene Wigner, a former schoolmate of von Neumann at the Lutheran Gymnasium in Budapest, began dividing his time between Princeton and the University of Berlin in 1930. In 1933, his post in Berlin was terminated due to his Jewish background, despite the fact that he and his entire family had converted to Lutheranism when he was a teenager.

With backing from the Rockefeller and Carnegie Foundations, the New York-based Emergency Committee for the Aid of Displaced German Scholars brought close to 300 scholars out of Germany to safety. A number of German-speaking scholars wound up in Princeton. In 1935, a young American mathematician, Leonard Blumenthal, secured an appointment as assistant to Weyl at the Institute for Advanced Study. Blumenthal's wife Eleanor wrote to Frieda, "Princeton is full of foreigners. I can't imagine that there are any Germans left in their native land. In fact, at every gathering, more German is spoken than English."[8]

While the Institute for Advanced Study snapped up some of the best and brightest mathematical stars fleeing Hitler's regime, the administration at Chicago allowed several opportunities to pass it by. Finally, with Dickson's retirement looming, Adrian proposed a candidate he believed would pass muster with a Department that included but one token Jewish member—Adrian himself. He nominated a former student of Noether at Göttingen, Otto Schilling, who had managed to secure a somewhat grudging letter of recommendation from Hasse.[9] Schilling, the son of a well-to-do Aryan German family, became Hasse's student when Noether was forced to leave Göttingen. Schilling sailed through the appointment process and joined Chicago's faculty in 1939.

Adrian was happy to have a disciple of Noether and Hasse on the faculty and bent over backwards to make Schilling and his fiancé feel welcome. The Alberts invited the couple to a private, get-acquainted dinner at their home. The following month, they threw a party to introduce the newcomers to members of the Department.[10] Upon returning from the party, the bride-to-be wrote to Frieda, "I can't go to bed tonight without telling you and Adrian how happy you made Otto and me today. It was such a beautiful party."

As matters in Europe worsened, some of the refugees worried about becoming nationless. Brauer settled into teaching at the University of Toronto. In September 1939 Adrian wrote from the American Mathematical Society Colloquium in Wisconsin, "Brauer was afraid to come. He isn't a British subject yet and thought that if he left Canada he might not get back."[11]

President Franklin Delano Roosevelt was inching toward supporting the Allies in Europe. The King and Queen of England paid a state visit to Washington, D. C. in June 1939. The popular couple sat in an open car waving to an adoring crowd of well wishers lining Constitution Avenue. Afterwards, Eleanor and Franklin Roosevelt entertained King George VI and Queen Elizabeth at a state dinner in the White House, where they were treated to the breathtaking tones of divas Marian Anderson and Kate Smith.

The following month, President Roosevelt initiated a series of steps leading up to America's involvement in the war overseas. He asked Congress to repeal the arms embargo that prohibited the United States from selling U.S. arms to countries such as England, and to revise the neutrality law. He followed up by instructing his Secretary of State to abrogate the trade treaty between the U.S. and Japan signed in 1911.[12]

Some observers clung to the belief that all-out war could be avoided, especially when Germany and the U.S.S.R. signed a mutual non-aggression pact in late August. However, the die was cast when Germany invaded Poland on the first day in September. After Germany refused to withdraw, Britain and France formally declared war against Germany. World War II had officially begun in Europe.

On a Saturday morning in mid-July 1939, Einstein served tea to Szilard on the screened porch of his summer cabin in Long Island. It was not merely a social visit. Szilard had come to enlist Einstein's aid in sounding an alarm to President Roosevelt.[13]

Szilard and Wigner had received reliable information that German scientists were developing the deadliest of weapons of mass destruction. Fearing the unthinkable, the two scientists drafted a letter to Roosevelt about Germany's stockpiling of uranium that could be used to set up a nuclear chain reaction. They wanted Einstein's imprimatur on the letter to give it credence. Einstein agreed that the gathering threat was real, and signed it.[14]

Szilard did not entrust Einstein's letter to the U.S. mail; instead, he gave it to a confidante of Roosevelt, economist Alexander Sachs, to deliver personally. Although Sachs received the Einstein letter in early August, Roosevelt was too busy to meet with Sachs until October. At their October meeting, Sachs handed Roosevelt Einstein's letter along with an explanatory memorandum prepared by Szilard. Roosevelt immediately grasped what was at stake. "Alex," said F. D. R. to Sachs, "what you are after is to see that the Nazis don't blow us up."[15]

In January 1939, Adrian received an invitation to deliver a "stated address" at a Conference on Algebra scheduled as part of an International Congress of Mathematicians. The Congress of 1940 was set to convene in Cambridge, Massachusetts. However, the war in Europe caused the international group to make a sudden change in plans. He notified Frieda from the Colloquium in Wisconsin, where he heard the news. "The Congress is postponed indefinitely. What a life! Now I'll have to make another algebra conference next summer."[16]

The 1940s ushered in tense times for all Americans. On the other hand, the forties signaled an end to the Great Depression. Women joined the work force in record numbers. Factories were humming. The nation pulled together in support of the war effort.

Adrian's career advanced. He received tenure and a full professorship in 1941 at the age of 35. In those days, universities rarely granted full professorships to anyone under the age of 40.[17]

His full professorship was not the only unusual thing about the young mathematician—his teaching style provoked comment as well. According to a profile in the *Chicago Daily Tribune*, Professor Albert "often illustrates his lectures by rapidly writing the formulas on the blackboard. With the chalk in his left hand he fills the left half of the blackboard with figures and formula beginnings, then switches the chalk to his right hand and continues them without missing a figure."[18]

His personality had a whimsical side. Someone once asked him, "What do mathematicians *do*?" He replied, "*Here's* what they do." He quickly wrote two long columns of five-digit numbers on the blackboard. Then he added up the left and right-hand columns simultaneously, using both hands.[19]

Although the United States did not officially enter World War II until after Japan's attack on Pearl Harbor in December 1941, the University of Chicago began mobilizing for war long beforehand. In the February 28, 1941 issue of the *Anniversary Times* celebrating the University's 50th anniversary, the headline blared, "'All Out' for U.S.—U. of C."

The entire second page of the issue was devoted to the University's all-out defense effort. Almost every department provided special services to help in winning the war. One of the "special services" highlighted in the article was military cryptography. Adrian already served as a consultant in cryptanalysis for the War Preparedness Committee of the American Mathematical Society by then, having begun this work in 1940.[20]

During World War II, America played a high stakes game of hide and seek with the enemy with respect to military plans and troop movements. Cryptography and cryptanalysis were two of the most critical battlefields during the war. Adrian applied himself to fighting on both fronts.

American scientists and military personnel used "cryptography," the science of codemaking, for devising devious codes to baffle the enemy. At the same time, they employed "cryptanalysis," or codebreaking, to crack the enemies' codes. The broader term, "cryptology," encompasses the scientific study of both cryptography and cryptanalysis. The unwieldy terms "cryptography," "cryptanalysis," and "cryptology" all stem from the Greek word "kryptos," meaning, "hidden."[21]

In September 1941, Adrian chaired a Conference on Algebra at the University of Chicago, where the American Mathematical Society held its 23rd annual Colloquium. The 5-day Conference coincided with an extensive science program planned to commemorate the University's 50th anniversary.

Several luminaries from the 1938 Conference on Algebra spoke at the 1941 meetings and a number of other mathematical stars enhanced the event. Marshall Stone arrived from Harvard to deliver a one-hour address on Abstract Spaces. Gilbert Bliss received equal time for his lecture on the Calculus of Variations, a relevant topic given the war that was raging in Europe. Solomon Lefschetz spoke on the latest advances in topology. Others such as John von Neumann, Oscar Zariski, and Nathan Jacobson rounded out the program with some shorter lectures on cutting edge mathematics.

Based on his pitch that a number of VIPs would be present, Adrian persuaded the University to underwrite a gala banquet at a Hyde Park hotel. Eleanor Blumenthal sent a note to Frieda afterwards, thanking her for the "elegant dinner party" held in conjunction with the conference. The note was sealed in an envelope bearing three one-cent stamps with a picture of Lady Liberty above the legend "FOR DEFENSE." Mrs. Blumenthal added, "We are counting on Adrian giving his 'code' talk and coming our way en route."[22]

It was two weeks before Pearl Harbor when A.A. Albert spoke at a meeting of the American Mathematical Society. He delivered a groundbreaking lecture about what had been, until then, the slowly evolving science of cryptography. A young mathematician named Lester Hill began experimenting with using algebra as a process for cryptography in 1929.[23] During the course of his work for the War Preparedness Committee, Adrian reviewed the rather preliminary work of Hill and others and concluded that all of the cryptographic methods then in use were special cases of *"algebraic* cipher systems."[24]

Adrian may have been the first to recognize the potential advantages of applying the techniques of *abstract* algebra to cryptography.[25] In the address, he presented his novel conclusion that "cryptography is more than a subject permitting mathematical formulation, for indeed it would not be an exaggeration to state that *abstract* cryptography is *identical* with abstract mathematics."[26]

He went on to describe how the use of modern algebraic techniques made it possible to progress from primitive cipher systems to sophisticated ones. His analysis made use of the fundamental concepts of modern algebra—those of ring, finite field, and matrix. Adrian's reformulation of cipher systems in algebraic terms made previously unsolvable problems in cryptography solvable.[27]

Adrian approached cryptography in the same fashion that he attacked the problems of associative and nonassociative algebras. He reviewed all previous work in the field and then began to classify it.

Shortly after delivering the address, he polished his lecture in the paper "Some Mathematical Aspects of Cryptography." He sent the paper to the War Department and the Navy, but never attempted to publish it. Most likely, it was due to the secrecy of the topic that the paper lay unpublished for fifty years until the American Mathematical Society included it in a two-volume collection of his manuscripts.

At Chicago, Professor Albert marshaled a group of handpicked graduate mathematics students to assist him in his defense work on cryptography. According to the school newspaper, the *Daily Maroon*,

> As problems arise in connection with the government's work on cryptology they are sent to Mr. Albert for solution—if possible. The cryptographic methods used in World War I were very primitive compared to the methods now used. Mr. Albert suggests the present day complexity of this science when he said, "I have discovered that the field of my own research in algebra has applications in cryptology—the enciphering of messages—and I have devised some new cryptographic methods which I hope may be of service in the military intelligence field." The present day mathematical codes will be much more difficult to break than any known at the time of the last war.[28]

The *Daily Maroon* goes on to point out that Albert's 1941 book, *Introduction to Algebraic Theories*, provides the background in matrix theory used to understand the new cryptographic methods.

Scientific advancements in the field of cryptology remained top-secret long after the end of World War II. In 1969, Adrian received an inquiry from a Massachusetts professor who wanted to teach a class in the subject. The professor asked how group theory applied to cryptography. Adrian replied that he did not know of applications "other than in material which is classified...." He referred the writer to references contained "in a huge book called the *Codebreakers* by David Kahn." He continued, "The book is not completely accurate but is fascinating...."[29]

While still serving as an editor of the *Bulletin*, he received a copy of a letter circulated by Dean Fort seeking reaction from editorial board members to a proposal coming from an American mathematician. The mathematician recommended that American journals reply in kind to German publications by refusing to recognize papers by German mathematicians. Adrian replied:

> ...The suggestion made that American journals adopt the policy of giving no reference to German periodicals

seems to me to be utterly fantastic. I don't know who this was intended to annoy but I am sure that the Nazis would not be hurt even slightly by it. It would undoubtedly result in increased difficulty in learning mathematics by Americans if it were carried out. If it were extended in a logical fashion all "German" mathematics should be ignored and the statement that there is such a thing as "German" mathematics is just as stupid as the corresponding statement in *Deutsche Mathematik*. If we cannot conduct our mathematical work without emulating precisely those tactics of the Nazis which we so wholeheartedly condemn, I for one feel we ought to quit our jobs and do something more worthwhile. I am afraid that...[the American professor] seems to have missed the point of what it is we are fighting about and I hope that his suggestion, which is the normal and well-deserved hot-headed reaction to Nazi tactics, will be ignored. I am sure that he himself would regret such a policy after this mess we are in is cleared up.[30]

In the midst of the war, A.A. Albert was elected to membership in the National Academy of Sciences. From the time Abraham Lincoln signed its charter in 1863, the Academy played a unique role in American science. Incorporated as a private, non-profit society, the Academy's charter required that it, "whenever called upon by any department of the Government, investigate, examine, experiment, and report upon any subject of science or art..."[31] The June 1943 issue of the Academy's *Scientific Monthly* featured an article by the President of the Academy titled "National Academy of Sciences and Its Part in the War—the Academy in Relation to the Government."

Like many other American scientists, Adrian pitched in to help the war effort in a variety of ways. In November 1943 the University appointed him "to give extra service in the Department of Mathematics and for the Army Specialized Training Program."

The advent of World War II shook American mathematicians out of their ivory towers. Their notion that a sharp dichotomy existed between pure and applied mathematics suddenly softened.

President Roosevelt established the Office of Scientific Research and Development (OSRD) to coordinate all government-sponsored scientific efforts.[32] Through OSRD, the government contracted with universities around the country to conduct research on behalf of the military. Several universities received funding as part of the OSRD's Applied Mathematics Panel. By placing pure mathematicians on the Panel's advisory board, the government hoped to entice their fellow academicians to work for the military establishment. Adrian knew some of the advisory board members personally, as the University of Chicago's Lawrence Graves and Princeton's Oswald Veblen and Marston Morse were among them. He was also friendly with the technical aide at OSRD, Mina Rees, a former doctoral student of L. E. Dickson.

The government's well-considered approach was a great success. Pure mathematicians joined the ranks of OSRD projects in droves and regarded the assignments as plums. In the fall of 1944, Adrian began consulting as a Research Mathematician for the Applied Mathematics Panel at Northwestern University. This necessitated driving from his home on Chicago's south side clear across the city to the North Shore suburb of Evanston at a time when gasoline was rationed. Chicago granted him a leave of absence for most of the following year so that he could serve as Associate Director of the group at Northwestern.

Mayme Logsdon, one of Adrian's former mathematics professors, left Chicago after reaching the University's mandatory retirement age. Upon learning of his appointment at the Northwestern war project, she wrote to her pupil from her new post at a small southern college:

> ...[D]ouble congratulations (a) on the appointment on the Applied Mathematics panel and (b) on the completion of the...[book on] College Algebra. Who is publishing it? It can't be ready by Autumn quarter surely. Put me down for 102 [copies] in the Spring and I'll use it as my text....
>
> These Navy boys are all just out of High School and do not do as good work on the average as the ASTP boys. My advanced classes have been pushed too fast and [they] have not learned well what they have been exposed to. It is discouraging....Will you have to go to Evanston each day this Autumn [to work on the applied mathematics panel] or can you do most of your work in your office? You can get extra gasoline for necessary trips, I am sure.[33]

Logsdon was not alone in observing that the nation's military was unprepared. At the beginning of the war, General Lewis B. Hershey, head of the United States Selective Service System, openly criticized the woeful weakness of America's youth in mathematics.[34]

Adrian attributed the problem to neglect of the subject in America's school systems. He described high school mathematics curricula during the war as "an illogical mess of seemingly unrelated items."[35] To add insult to injury, some states allowed students to graduate from state-run universities without ever having taken a mathematics course beyond the grade school level.[36]

He believed the architects of the policy to be some powerful educators headed, in the 1940s, by the Teachers College of Columbia University. The group made a concerted effort to remove mathematics from the list of requirements for college entrance.[37] In his view, the Teachers College was largely responsible for the emphasis on teaching teachers more about training methods than about content. He charged, "The Teachers College edict that anyone with proper teacher training could teach anything is one of the great frauds of our time."[38] University professors like Adrian began to address the problem during World War II by revamping curricula, rewriting inadequate texts, and serving as advisors to colleges seeking to upgrade their mathematics programs.

His skill-set included one tool considered unusual for a mathematician—managerial ability. This came in handy at the Northwestern University war project, where matters were in disarray when he assumed some administrative functions.[39] The staff was becoming increasingly distraught over seemingly contradictory orders conveyed by various generals. The civilian professor from Chicago called all of the generals in to attend a meeting where he solicited their input and then gave a long explanation about the project. Afterwards, the apparent discord disappeared and things functioned smoothly.[40]

Saunders Mac Lane headed a similar group at Columbia University's Division of War Research. He had urged Adrian to join his group, explaining, "Our work is not sophisticated in the mathematical sense; in fact, spherical trigonometry and linear differential equations constitute a good portion of our technical stock in trade. Nevertheless, our problems do arise from real military situations...." When the Chicagoan declined, Mac Lane replied with an understanding letter regarding Adrian's decision to choose the project at Northwestern instead, noting how difficult it is to move a family.[41]

A member of Mac Lane's group later gave a more explicit description of the work at Columbia:

The problem that we were presented with as a group was the problem of plotting the course of an attacking fighter against a bomber. The problem seemed to be this. The bombers flew straight and level at a consistent speed, and they had guns with handlebars that could be aimed. The fighters flew any old way with the guns hitched in their wings; in order to aim the guns you had to aim the airplanes. To tell the gunners on the fighters where to point their guns you had to know where to expect this crazy plane to be. The existing doctrine was wrong, and as a result we lost a number of planes, I think in the African campaign. So it was a matter of some importance to get what was called an aerodynamic attack course, so you could figure out where the planes were.[42]

The Northwestern and Columbia projects, both sponsored by OSRD, worked in concert. Adrian spent so much time at the Columbia project that some researchers there assumed he was part of its staff.

During his visits to the OSRD project at Columbia, he mentored two promising young mathematicians, Daniel Zelinsky and Irving ("Kap") Kaplansky. Zelinsky, one of Adrian's prize doctoral students, had interrupted his graduate studies to aid in the war effort.

Kap received his doctorate from Harvard in 1941, three years after the Algebra Conference in Chicago that had "put a stamp" on his career. He had the distinction of being Saunders Mac Lane's very first Ph.D. student. After graduation, he was appointed a Benjamin Peirce instructor at Harvard in 1941 and kept the post until joining the OSRD project in 1944.[43]

Kap later expressed his appreciation for the Chicagoan's patience in guiding him through the fine points of aerial photography. He recalled, "I don't remember how I became aware that Adrian was looking at aerial photography. But I did, got an idea, and communicated it. He was a big shot, I was at the bottom of the ladder, but he was kind to me."[44]

Although the pure mathematicians had no previous experience in applied mathematics, their analytical skills made up for their lack of on-the-job training. Adrian later commented that one of the pure mathematicians at Columbia managed to solve a problem about fighter plane pursuit curves that stymied the applied mathematicians.

Another University of Chicago mathematics professor, Magnus Hestenes, came away from the Columbia project with a similar observation:

> I found that my war experience gave me a broader outlook on the role of mathematics in society. It also gave me a greater appreciation of research mathematicians. A good researcher, no matter what field, would tackle a problem with vim, vigor and imagination. He needed no guidance. This was not true for some mathematicians who were not researchers although they did good work with guidance.
>
> In my latter years I asked myself what branch of mathematics was most useful in applications. I came to the conclusion that it was algebra. Accordingly, algebra can be viewed to be the basic course in applied mathematics. I doubt if many algebraists view algebra as applied mathematics.[45]

Frieda did her part in the war effort, too. She helped raise money for the Red Cross. Thanks to many like her, Chicago surpassed its Red Cross War Fund quota of $11 million. Hyde Park raised an even higher proportion of those funds than the rest of the city.

Frieda's parents and four of her siblings had relocated to California, which was on high alert during the war. Military officials assumed that the west coast would be the first to suffer if the mainland were invaded. One night, Los Angeles awakened to the sound of antiaircraft weapons. It proved to be a false alarm—although U.S. weapons were fired, the military never confirmed an enemy attack.

During the war, Frieda was constantly glued to the radio. Uncle Sam drafted her younger brother Hank into the army even before Pearl Harbor. Hank's new home became an underground bunker where he learned to fire antiaircraft weapons.

Not long afterwards, her youngest brother, Danny, was drafted as well. Danny's draft board placed him in the Air Force. Although his vision prevented him from piloting a plane, he held a high-risk job as a mechanic seated in the cockpit with the pilot and copilot.

With her brothers in the war and the future so uncertain, Frieda's only sister, Ethel, eschewed a formal wedding. She married her beau in a simple ceremony and telegraphed the news to Frieda afterwards.

Hank wrote a series of letters designed to comfort Frieda. The letters advised her not to worry about the soldiers' lack of sleep—it was no longer a problem because they had grown used to it. Or the bad food—it was no problem because everyone had adjusted to it. In another letter, he drew her a diagram of their underground bunker complete with submachine guns to demonstrate how safe they were.

Frieda was beside herself with worry. In September 1943 she flew to California to comfort her parents—and herself. Adrian was obliged to do his own work and care for three young children in her absence. He maintained a steady barrage of love letters designed to hasten her return to Chicago. In one such letter he remarked,

> The butcher said he would never take a job like I have and everyone else has been kidding me. They don't realize how much I love you. Men have fought wars for women. I'm only cooking, cleaning and washing dishes and I'm sure no one ever loved a woman as much as I do you. xxxx
> I feel happy just thinking what a lot of joy you must be experiencing and it makes up for my loneliness.[46]

One of Frieda's cousins serving in the Army wrote from overseas in 1944, "Are there many men in the math department at the U. of C.? In London, there are very few men in the university because most everyone's in the service."[47]

About a year after America entered the war, General Hershey issued a bulletin certifying that college mathematics instructors and professors in America were essential to the war effort. He urged them to apply for deferments. Nonetheless, America's store of Ph.D. mathematicians dwindled, as many undergraduates were unable to complete their education.

In December 1944 Chicago's former Mathematics Department chair Gilbert Bliss, by then aging and in ill health, wrote to Adrian, "I am hoping that the tide will turn again in our favor soon and release our boys, including the mathematicians, to pursue their normal activities again."[48]

Rationing was the order of the day. The Office of Price Administration issued a war ration book for each member of our family. The Albert family often ran out of "points" and had trouble getting eggs. In a letter to Frieda before her return from California, Adrian reported, "I went shopping today with all the blue points.

Canned goods, especially peas, have been conspicuous by their absence...I have been stocking up on soap when I see it as they only sell 2 bars to a customer."[49]

On the other hand, the Alberts were among the lucky ones. Our family members serving abroad lived with danger and harsh living conditions as well as rationing. Adrian's nephew, an Army paramedic, wrote from "Somewhere in England, 8 August 1944":

> Everything is strictly rationed and so money doesn't mean too much for you can only get your ration (if you're lucky) and that's all...Really amazing some of the sights, and the ruins of the bombings, the appearance of the people all bring out the rough time they have had in this war. The drone of plane motors overhead is quite the common thing now....[50]

Frieda's cousin reported from France the following month:

> ...I've been very busy chasing Adolph's gang back to the Fatherland....There are many things I would like to tell you but am unable to. I've been through a lot of fighting....We're on the go day and night. You can see by your newspaper what's going on.[51]

Even the University's landscape changed with the advent of the war. Portions of the placid Midway were now dotted with hurriedly constructed Quonset huts serving as temporary abodes for those involved in the war effort.

A top secret OSRD project occupied Eckhart Hall, and Mathematics Department personnel relocated to the library temporarily.[52] The research effort was dubbed "Manhattan Project" because it was administered by a section of the army code-named "Manhattan District."[53] For security reasons, Arthur Compton, Dean of Physical Sciences at Chicago, referred to the endeavor as "the metallurgical project at the University of Chicago,"[54] leaving the impression that it was nothing more than a wartime metals program.

The Project spilled over into the Jones Chemical Laboratory and the hidden underbelly of Stagg Field, where scientists were secretly constructing a graphite pile for use as a nuclear reactor. Indeed, along with many private universities during World War II, Chicago gave over a significant portion of its resources to research performed for the nation's military establishment.

The extraordinary collaboration between government, science, and academe in America's Manhattan Project was critical to the project's success. Across the Atlantic, Germany, with its government-controlled university system, failed to achieve a coordinated effort to advance a nuclear fission program.[55]

Werner Heisenberg, one of Germany's top scientists, had invented matrix mechanics, the first formalization of quantum mechanics, in 1925.[56] With the help of Pascual Jordan, who provided the mathematical framework for the new science, Heisenberg created quantum mechanics—a feat for which Heisenberg received the 1932 Nobel Prize in physics.

This should have given Germany the lead in developing nuclear fission. Hitler placed Heisenberg in charge of Germany's atomic weapons project, but gave the project tepid support because he deemed Heisenberg's physics to be "too Jewish" despite the fact that Heisenberg was of Aryan descent.[57]

From his Metallurgical Laboratory in Chicago, Compton, whose connections with prominent scientists were critical to organizing the atomic bomb project, coordinated the research conducted at Chicago and at other institutions including the University of California at Berkeley, Columbia University, the University of Minnesota, Princeton University, and the National Bureau of Standards.[58]

Although A.A. Albert never participated in the Manhattan Project, many of his colleagues and acquaintances did. Eugene Wigner, one of the trio who had asked him for help in identifying an exceptional algebra of quantum mechanics at Princeton, became one of the key physicists working at the Manhattan Project in Chicago after Compton ordered key personnel to gather at the "Met Lab" in the spring of 1942.

Wigner worked on the project under the direction of Enrico Fermi, who fled Italy in 1938 when anti-Semitism became the official policy of Benito Mussolini's Fascist regime. Fermi was a Catholic, but his wife, Laura, was Jewish. On his way into exile, Fermi had stopped in Sweden to receive the Nobel Prize for his work on the bombardment of atomic particles.

Laura Fermi, an accomplished scholar in her own right, later described the interactions of three European émigrés who played a major role in giving birth to nuclear fission:

> ...[Fermi], Szilard, and Wigner formed an exceptional team: Szilard threw out ideas; with his practical intuition Fermi turned them into rough theories that served immediately to guide experiments; and Wigner, more patient and rigorous, refined them into mathematically cogent theories that would stand the test of time.[59]

On December 2, 1942, Manhattan Project scientists quietly succeeded in achieving the world's first controlled nuclear chain reaction in a squash court beneath the bleachers of the University of Chicago's former football stadium.

Many of the Manhattan Project scientists who helped to create the bomb believed it was needed as a deterrent to Nazi aggression, but fervently hoped that America would never have to use it. In the spring of 1945, a number of American scientists became increasingly concerned about America's potential use of the bomb. Just as Germany was on the brink of surrendering to the Allies, American intelligence learned that German scientists had given up their efforts to build an atomic weapon.[60] There was no longer any rationale for using the bomb against the Nazis. Szilard began frantic efforts to lobby against use of the bomb.

Meanwhile, hostilities on the eastern front showed no signs of letting up. The War Department estimated that an enormous number of Americans would die if they attempted to land on the Japanese Islands. Estimates of American casualties ranged from 400,000 to one million.[61] The pain of the D-Day disaster on Omaha beach was fresh in the American psyche, and America could not tolerate a slaughter of American servicemen that might dwarf the D-Day catastrophe. In July, America and Britain gave Japan an ultimatum—surrender unconditionally, or else.

After several years of concerted effort, American scientists had managed to produce just two atomic bombs. Strategists debated a variety of alternatives for using the scarce weapons to terminate the war, including staging a public demonstration on a deserted island to prove their effectiveness. The problem with that option was that the bombs could misfire, making the demonstration counterproductive. In the end, the War Department concluded that, given the small number of bombs available and their unreliability, the option most likely to bring the war to a speedy conclusion would be a deadly demonstration of the bomb's effectiveness.

In August, the devastating power of America's secret weapon was revealed for all the world to see. The United States Air Force dropped one atomic bomb on Hiroshima and a second, on Nagasaki, annihilating untold thousands of Japanese civilians. Even the bombs' inventors were amazed at the enormity of their destructive power.

Einstein expressed shock and horror when he heard the news. Few in the press at the time questioned President Truman's decision to deploy the bomb. Most Americans were war-weary and joyful that the long and painful campaign had ended. President Truman later assured the nation that the sacrifice had been "urgent and necessary."[62] He believed that the bombing of Hiroshima and Nagasaki had actually saved tens of thousands of lives on both sides.

Once the war was over, the mathematics faculty at Chicago looked forward to seeing the military and its project leave the campus. Adrian wrote to Lefschetz in the early part of 1946, "The Army is reported in the process of evacuating Eckhart Hall and we hope that they will be out by the middle of April. There are rumors that we will return by July 1."[63]

Chapter 14

South of the Border

During the war Zelinsky, who loved the violin almost as much as he loved mathematics, met a beautiful young New Yorker named Zelda. Coming from a musical family, Zelda shared his affinity for music. Zelinsky was not your average beau—he broke the mold when he brought his violin over to accompany members of her family.[1] Before long, the musical mathematician had fiddled his way into her heart.

At the end of the war, government-sponsored research projects around the country were dismantled and mathematicians resumed their peacetime activities. As he packed up to leave the Columbia research project, the tall, lanky Zelinsky turned to his roommate, the athletic-looking bachelor, Kaplansky. "Zelda and I are getting married." Instead of flashing his customary dimpled smile, Kap looked chagrined. "Don't you know you'll get tired of any woman after about six months?"

Zelinsky found the comment somewhat disconcerting. When he returned to complete his studies at Chicago, Zelinsky repeated the comment to his doctoral advisor. Adrian assured him, "Pooh, pooh. He doesn't know you can love one woman the rest of your life."

The Zelinskys were surprised to encounter Kap on the train headed for Chicago in 1945. Kap had neglected to mention his own postwar plans. Harvard issued no invitation for him to return after his "strictly temporary" position as a Benjamin Peirce Instructor.[2] Fortunately, due to Adrian's machinations, the University of Chicago offered Kap an instructorship that would lead to tenure. The addition of Kaplansky indicated a slight thaw in the Mathematics Department's unofficial freeze on hiring Jewish faculty members. On this occasion, Chicago had also selected an exceptionally gifted mathematician.

Adrian Albert's conventional perspective on love and marriage was but one example of his straight arrow persona—a persona displaying none of the eccentricities of a stereotypical genius. The origins of his character lay in his background—a stable, supportive home life during childhood, midwestern values, and an upbringing steeped in Jewish tradition. It was character as much as mathematical prowess that powered his career.

The Chicago professor was 100% dependable. If he had an appointment, he kept it, on time. His mild demeanor seldom varied. If a colleague or student needed a favor, you could count on him to oblige. His faithfulness to his wife was unflagging. His strict code of ethics and personal behavior, though not the stuff of gossip columns or cocktail party chatter, no doubt enhanced his success at winning key policy-making positions and securing top-secret clearance.

Adrian dressed the part of a university professor. His standard office garb was a suit with a vest, which belied his expanding waistline. Wearing his characteristic bow tie and gentle smile beneath a high dome of a forehead, the bespectacled professor bore an uncanny resemblance to his former teacher, E. H. Moore.

Around the Quadrangles, folks saw him as a friendly man.[3] If he spotted someone he knew while walking to or from campus, he would wave, cross over, and chat for a minute or two, sharing his wit about the light sides of life.[4]

The fondness for gizmos that he developed as a youth continued into adulthood. He took a childlike delight in sharing the latest gadgetry with friends and colleagues.

Two things made him different. He had the heart of a marshmallow and the determination of a nuclear missile. Perhaps his empathy for "the little guy" came from his humble beginnings. He treated everyone, from the lowliest janitor to his superiors, in the same fashion. The maintenance crew was not accustomed to being treated with such respect. As a result, if Mr. Albert wanted a favor from the Buildings and Grounds Department, they sent someone over, pronto.

The University administration knew that if you wanted to get something done, "call Albert." Whether it was his own research, the Departmental budget, or writing a letter of recommendation for someone, he was relentless. Everything he had to do got done, and got done fast.

Adrian was a "doer," not a worrier. Frieda did enough worrying for both of them. He advised her, "If there's something you can do about a problem, then do it. Otherwise, don't worry about it."

On the surface, his seemed a tranquil life. But there were troubles bubbling beneath the surface. There were stresses brought on by concerns about his young diabetic son and by the war. There were the usual conflicts with the University's administration.

Most stressful of all was his internal conflict each time he realized that one of his doctoral candidates did not have what it took to earn a doctorate in mathematics. Inevitably, he became so heartbroken that his digestive system went into turmoil. Having dreamed the dream himself, he knew how much it meant to pursue a career in mathematics. Worrying about a problem and agonizing over it were two different matters.

Doctors advised him to treat his gastrointestinal distress with a diet devoid of fiber. The prescription seemed to work, but a diet rich in simple carbohydrates caused him to gain weight.

Adrian apparently felt quite comfortable discussing even the most personal matters with Lefschetz. He wrote, "Did I tell you my colitis has completely disappeared? I certainly did get stout, though."[5] The weight gain, compounded by lack of exercise on account of his ill health, contributed to his next affliction. In his forties, he contracted adult-onset, or type 2, diabetes.

In the summer of 1946, he became Acting Chairman of Chicago's Mathematics Department for the second time. That year, he published his fourth book, *College Algebra*.

Mathematicians worldwide breathed a collective sigh of relief now that the war was over and travel became easier. The University of Chicago held its Third Conference on Algebra in July. Chicago's three algebraists, Adrian, Kaplansky, and Schilling, were among the lecturers. Other speakers included: Harvard's Stone, Mac Lane, and Garrett Birkhoff; Toronto's Richard Brauer; Reinhold Baer and Zariski (both then at the University of Illinois); Wisconsin's Richard Bruck; Jacobson (then at Johns Hopkins); Samuel Eilenberg (then at New Mexico); and Irving Segal (then at the Institute for Advanced Study).

Princeton followed up by convening a grand December celebration in honor of the 200th birthday of its predecessor, the College of New

Jersey. Among the festivities was a Bicentennial Conference on "Problems of Mathematics." The *Princeton Herald* reported that among the one hundred leading mathematicians and other scientists participating in the conference were Emil Artin, Joseph H.M. Wedderburn, Solomon Lefschetz, Nobel Prize winner Albert Einstein, Oswald Veblen, Hermann Weyl, John von Neumann, Richard Brauer, Oscar Zariski, A. Adrian Albert, and scientists from eight foreign countries.[6]

After returning from an exciting year in Brazil with his wife, Zariski nominated Adrian for a visiting professorship in South America. He confided to Lefschetz, who was fond of travel, "I hope that matters can be arranged as I have the wanderlust and am anxious to go."[7] The University of Chicago professor was slated to spend half a calendar year as Visiting Professor at the University of Brazil* in Rio de Janeiro, followed by about a month at the University of Buenos Aires in Argentina—both through the auspices of the U.S. State Department's International Exchange program.

There were many preparations to be made. The entire family underwent a series of immunizations. Everyone came through them fine, except 15-year old Alan, who fainted during the yellow fever shot. Then came the passport photos. And the packing. In addition to her pots and pans, Frieda brought along a year's supply of powdered milk. In 1947, one could not buy pasteurized milk in Brazil.

The Albert family drove from Chicago to New Orleans to have our car shipped to Brazil. We arrived in New Orleans on February 5, 1947 to the disheartening news that our sailing had been cancelled. We were obliged to await the next sailing, whenever that might be.

The delay afforded plenty of time to tour New Orleans and Baton Rouge. Tulane University declared a two-day holiday in honor of the Mardi Gras. I thrilled to the sights and sounds of my first Mardi Gras, with its magnificent floats. We gladly accepted an invitation for Roy and me to attend a scavenger hunt at the home of Professor Parker, a mathematician at Louisiana State University in Baton Rouge. Only Adrian had to work meanwhile, lecturing at Tulane University and LSU.

---

* The University of Brazil has since changed its name to the Federal University of Rio de Janeiro (Universidade Federal do Rio de Janeiro—UFRJ).

Frieda observed in her diary that the Louisiana State Capitol was probably the most beautiful in the country, and very historical. Leading to the formal garden were 13 steps representing the original colonies and then, 36 steps, one for each of the other states in the country with the year of their admittance to the union. The top step bore the words, "E Pluribus Unum." She added, "Huey Long left a few good works."

Even for Adrian, the visit to Louisiana was not all work. The LSU and Tulane mathematical families did all they could to entertain the Alberts while our parents fretted over when the ship might sail.

We children found the family of our father's former classmate, W. L. ("Bill") Duren, particularly congenial. Their son, Peter, was the same age as Roy, and their daughter, Sally, my age. Frieda wrote, "They and Alan had a grand time playing basketball, football, etc. Nancy couldn't be torn away from Sally."

Professor and Mrs. Rickey of L.S.U. invited our parents to an evening bridge party. Adrian won first prize, a heart-shaped box of Whitman chocolates.

After cooling our heels for nearly two weeks, the Alberts finally set sail for Rio de Janeiro. We left the known for new sights and sounds—flying fish, conversations in Portuguese, and palm trees sighted while cruising the Caribbean.

Frieda's diary noted that we passed the equator on Sunday, March 2. She opined, "It looked just the same—King Neptune didn't arise to greet us." Nevertheless, the ship's captain memorialized the event by handing out gold-sealed certificates in the name of Neptunus Rex, Ruler of the Raging Main, who commanded that all living things of the sea such as mermaids, sea serpents, dolphins, pollywogs, and whales must abstain from maltreating the ship's passengers should any of us fall overboard.

Adrian cast his fishing line off the side of the ship and caught a pompano. He gave it to the ship's cook to serve for dinner.

The Albert family arrived in Brazil on Saturday, March 15. Frieda observed, "The harbor is as beautiful as they say...Sitting comfortably on deck watching the fog lifting over Sugar Loaf and Hunchback Mts. It's a beautiful view with the city snuggling with its skyscrapers in front of these majestic mountains."

One of our first residences upon arriving was the Riviera Hotel overlooking Rio de Janeiro's breathtaking Copacabana Beach. It was a temporary abode until we could move into something more suitable. Adrian photographed all three of us children on the sand holding a humongous beachball with Sugar Loaf Mountain looming in the background.

Eventually, we rented a comfortable, furnished house on Ruo Cupertina Durão. The back yard was graced by a beautiful poinsettia tree and a water well—which came in handy at times when the city's water pressure took a sudden plunge. The mattresses in the upstairs bedrooms were stuffed with straw. The homeowners had left us two other amenities—Rosa, the maid, who spoke no English, and a cat with a serious bowel disorder.

Rosa took care of cleaning the house. Frieda did the grocery shopping and made our meals. Meal preparation required boiling water long enough to kill the microbes in the tap water and mixing it with powdered milk.

Frieda gave us strict orders regarding hygiene: drink water only after it's boiled, and only eat fruit if you can peel it first. The fruit in Rio was delectable—there were varieties of bananas we had never seen before—"apple bananas" that tasted like apples and mouth-watering tiny "silver bananas."

The North American professor's wife had a bit of adjusting to do when it came to Brazilian customs. She could not understand why people sometimes stood at the gate of our house clapping their hands and making such a racket. It took some doing for Rosa to help her understand that hand clapping at the gate was the Brazilian equivalent of ringing the doorbell.

Our rented house sat on a block occupied by middle class homes. At the end of the street, the neighborhood took on a starkly altered aspect. There stood a favela—a community of tiny shacks with dirt floors. The shacks clustered around a communal water pump where the women washed their laundry and laid it out on stones to dry in the sizzling tropical sun.

Rio was a city of contrasts. There were the favelas. And there were the grand plazas with beautifully inlaid mosaic tiles and carefully manicured topiary.

Within days of our arrival, Frieda enrolled me at the Escola Americana—a private school run by an English couple who were University of Chicago graduates. The student body consisted of English-speaking foreigners and the offspring of well to do Brazilians.

The quality of the education I gleaned that school year surpassed that of my Chicago public school by a multiple of three. At Chicago's William H. Ray Public Grammar School, every child in the class studied the same curriculum at the same pace. When a child consistently performed above the level of the class as a whole, he or she would be double-promoted, thereby skipping the following semester. The net result was a faster, but not necessarily better, education.

I was one of about six students in my Ray School class double promoted the year before my family left for Brazil. When I arrived at my new school, the Principal faced a quandary about where to place me because Escola Americana had no midyear grades. Rather than reverse my promotion, he elected to advance me another semester. Frieda spent many an afternoon tutoring me at the dining room table until I caught up with the level of my class in mathematics.

It was not until I returned to Chicago the following year that I became aware of the relative mathematics level of my class at Escola Americana. Those bright private school students, who engaged in frequent math competitions with each other, had advanced not six, but a full eighteen months beyond the curriculum in my Chicago public school class.

Even the privileged children at Escola Americana were subject to the downside of living in an underdeveloped country. I watched in horror one day as a young Brazilian girl fell from a seesaw in the school playground. Her right forearm snapped like a twig. "Children who live here have brittle bones because the milk isn't safe to drink," Frieda told me afterwards.

In September, U.S. President Harry Truman arrived in Rio. He had traveled to Brazil for the signing of the Western Hemisphere Mutual Defense Pact at the Inter-American Defense Conference held in the resort city of Petrópolis, an hour's drive from Rio. U.S. citizens were invited to the United States Embassy in Rio to greet the Trumans. I was one of the little people allowed to stand within shouting distance of the President.[8] For many years afterwards, Frieda kept a photo of President Truman, me, and a sea of other well-wishers on the corner bookcase in our living room.

None of our family spoke the native language, although we children studied Portuguese in school. But in hotels and many other places, we found English-speaking personnel. And our parents held in their arsenal another universal language that helped them to communicate both on the ship coming over and with some of the university faculty. As the offspring of Jewish immigrants, they both spoke Yiddish. Soon, they had a number of local contacts.

It did not occur to me at the time that Adrian's familiarity with Yiddish facilitated his career in mathematics. Since Yiddish is a "High German" dialect, his childhood exposure to the language was helpful. He augmented his early training by taking a full year of German in college. Consequently, he could read the texts of the German scholars who were breaking ground in modern algebra during his college years. Many of his papers and books employed German symbols, no doubt a carryover from the works of German mathematicians.

Our parents were invited to dinner at a typical Brazilian mathematics professor's residence—a tiny, one bedroom apartment the young professor, Leopoldo Nachbin, shared with his wife and mother. "How spotless your apartment is," Frieda remarked. To which Nachbin's mother responded, "It's not an easy task with a place this small. Professors here are paid so little. I've been praying that, after meeting you, my son could have a better life."

Frieda was touched. Some of the Alberts' shipmates on the voyage to Brazil were well-to-do business executives who expressed outrage at the paltry wages paid to U.S. college professors. After meeting this Brazilian family, she realized just how well off the Alberts were by comparison.

Several years later, I learned that Adrian had arranged for Nachbin to come to the United States as a Visiting Professor. Being an enterprising man, the Brazilian academic developed a lucrative import-export business on the side.

Since "Alberto" is a popular given name in Latin countries, it was not surprising when a local newspaper misstated the North American visitor's surname as "Adrian." However, the banner on the article was apt. It referred to him as "Mr. Algebra."

A.A. Albert played two roles in South America that year—teacher and good will ambassador. He taught a class in Modern Higher Algebra at the University of Brazil, gave a seminar on the results of his own research, and delivered a paper at the Brazilian Academy of Sciences.[9] The classes at the University of Brazil went well, although they were disrupted by many religious and official holidays, some unofficial holidays, and a month long strike by university students.

The North American professor took his role as good will ambassador seriously. He lectured at the Instituto Brasil-Estados Unidos on "The College of the University of Chicago." And he actively recruited Brazilian mathematicians to join the American Mathematical Society and subscribe to AMS publications. When he arrived in Brazil, only three Brazilian mathematicians were members. Shortly after his arrival, he recruited ten new members in Rio and six more in São Paulo. In his report to the Division of International Exchange Persons, he apologized for "the small number of recruits," explaining it was due to the fact that he ran out of application blanks. He expected to bring the number up to 25 by the time he left.[10]

During his stay in Brazil, he delivered several lectures at the University of São Paulo. One of these dealt with "nonassociative rings," which he described as mathematical number systems arising in connection with the measurement of statistical quantities in quantum mechanics. "Of particular interest," he explained, "is the fact that quantum mechanics is the basic theory behind atomic energy calculations. Its properties are therefore of interest to the physicist as well as to the mathematician."[11]

According to one of his confidantes, he experienced a small difficulty while teaching in Rio. After waiting through several pay periods, none of his paychecks had materialized. He finally asked the Chairman of the Mathematics Department at the University of Brazil to look into the matter.

The Chairman discovered that a very high-ranking official of the University had deposited the paychecks into a bank account and was collecting the interest on them. Whereupon, the Chairman, in an emotional confrontation with the official at a faculty senate meeting shouted, "My honor is at stake. If you do not pay Albert his salary, I shall commit suicide." Afterwards, the salary checks began to appear on a regular basis.

Of greater moment was the sad news that reached him from Chicago. In a poignant note to Kline, he described it as follows:

> We have had sunny weather for a month, but yesterday, the day of the eclipse, it rained here all day. The weather was quite appropriate, however, as I received the news yesterday that my mother had passed away suddenly of a heart attack.

In October, we left Brazil for Buenos Aires. For the entire month of our stay, we were housed in a luxurious hotel. Those were the days when Argentina was still blessed with a surplus of cattle. At dinner, no matter what entrée one selected from the menu, the house added a sumptuous filet mignon, gratis.

Adrian lectured for three weeks at the University of Buenos Aires and in the Universities of Tucumán, Córdoba, Rosario, and La Plata. Afterwards, he lectured in Uraguay before we returned to Rio to board the ship that would carry us back to the States.

Chapter 15

Hyde Park in the Forties and Beyond

Spending the better part of a year away from the States had its consequences. Adrian Albert returned from Brazil to discover he was embroiled in a hornet's nest of controversy.

He had kept his post as Managing Editor of the *Transactions* while out of the country, believing he could manage the journal "by remote control"[1] from Brazil as he had done from Chicago. However, before leaving, he took the precaution of adding Kaplansky to the panel of Associate Editors[2] to pitch hit for him as needed.

Upon returning, it was clear the Managing Editor had the big picture well in mind, but suffered a disadvantage by being out of touch with the daily details of events transpiring during his absence.

Two young mathematicians at Princeton, Richard Bellman and Harold Shapiro, had written a paper addressing the problem of determining the probability that two numbers are "relatively prime" (that is, the numbers have no common divisors larger than 1). They submitted their manuscript to the *Transactions* while Adrian was away.

Kaplansky arranged for the brilliant, but eccentric, Paul Erdős to referee the paper, as Erdős had a reputation for exceptional insights into problems involving number theory and the prime numbers in particular. Erdős not only refereed the paper, but upstaged the two mathematicians by improving on their proof.[3]

Bellman and Shapiro spoke to Erdős about the matter, and they all agreed to have the original paper printed with a footnote stating that the referee had an alternate proof. However, after a further delay, Kaplansky returned their paper with a letter containing Erdős' shorter proof, and suggested a joint paper crediting all three mathematicians.

Bellman and Shapiro stood pat, demanding that their paper be accepted as originally presented.[4] At that point, Kaplansky agreed to print the original paper and sent it on to the AMS' New York office to be forwarded to the printer for typesetting.

Afterwards, Erdős made the rounds of math parties claiming that Bellman and Shapiro were hijacking his solution.[5] The experienced mathematician's "one-upmanship" in developing an elegant proof and his subsequent well-publicized boasting about it left Bellman and Shapiro feeling humiliated, and they wanted to put the entire scenario behind them.

By that time, A.A. Albert was back in Chicago, and they wrote directly to him: "we have decided to withdraw the paper completely. We are sorry to have delayed so long in making this decision...."[6]

Upon reading the retraction, Adrian immediately set in motion the procedures for withdrawing a paper. First, he notified the New York office. They in turn informed him that the paper had been sent to the printer in Wisconsin some weeks earlier. The AMS staff attempted to halt the typesetting, but it was too late, and they issued a final order to kill the type.[7]

If he had a personal failing, it was his tendency to see things in stark terms of right vs. wrong, occasionally losing sight of the human side of the equation. In his mind, Bellman was in the wrong, period. Postwar *Transactions* budgets were tight. They were receiving an ever-increasing number of deserving papers, and space was at a premium. AMS officials proposed an editorial policy requiring all authors to "ruthlessly compress" manuscripts and "avoid space-consuming exposition." Discussions were underway with two universities to explore whether one of them might start a new journal to pick up the slack.

Adrian was concerned lest a substantial backlog of papers awaiting publication accumulate as had happened at the *Bulletin*. And in his mind, the Bellman-Shapiro paper was a wasteful, "lengthy proof of results which could be obtained in a very few pages."[8]

Recalling that Dean Fort had billed two authors who retracted their paper after type was set, causing the *Bulletin* to incur an unnecessary printing bill, the issue seemed to be cut and dry. Nonetheless, he wrote to Secretary Kline for an official ruling on the matter.[9] Referencing Fort's action, Kline assured Adrian the AMS Council in that instance decided "editors have full power to assess charges in such cases" and opined, "I do not think that Bellman and Shapiro have any legitimate grievance."[10] As a result, a bill for $300 was sent to the authors for reimbursement of the cost of printing their paper.

When they saw the bill, the young mathematicians were in a state of shock and disbelief. They composed a second letter to the Managing Editor providing an explanation of events during the months he was away. They objected to the "drastic action" taken in destroying the type and insisted, "we do not feel that our letter in any way indicated that we wanted the paper withdrawn at any cost….we should have been consulted before this was done, if the editor intended us to pay for the cost of the type-setting."[11]

Whereupon, A.A. Albert vented his frustration in a rather intemperate letter delivering the ultimate generational rebuke to the two young mathematicians. Reminding them that their letter had explicitly and completely retracted the paper, he retorted, "I assumed you were adults who meant what they said…. It is my feeling that you have acted very badly in this matter."[12]

Upon receiving Albert's letter, Bellman went straight to Solomon Lefschetz to complain about his treatment. Knowing that Lefschetz was Bellman's Princeton mentor, Albert had already copied Lefschetz in on the correspondence.

Bellman, who later described the incident in his autobiography, imagined that Lefschetz intimidated Albert into acquiescing by bringing up some incident from Albert's past.[13] Such was not the case. Instead, Lefschetz reposed his considerable persuasive powers in a letter combining irrefutable logic with the wisdom of a senior statesman.

Lefschetz wrote: "Since you sent me a copy of your letter…I take it that you might like to know my reactions to the whole incident." He began by expressing sympathy for Adrian's predicament and attributing part of the problem to the logistics of having the journal's editors, AMS offices, and printing company in three different states. He then proceeded to chide his former protégé for impatience. Noting that he had reviewed all the correspondence and discussed the matter with "the two boys," Lefschetz declared:

> I…cannot believe that you really mean, as you put in your letter "that you had not done the best possible job in the mathematics of your paper". Certainly you would not say that if you had seen them working day by day. I must

remind you of what one of my colleagues told me once, that we editors are there to take it on the chin with patience.... [I]t is my feeling that authors are entitled to protection from referees and, in particular, against their possible tendency to keep "improving" even while the paper is "in course of publication." This is all the more true when, as in the present case, the authors are beginners and the referee is a top expert.... Frankly,...whenever one may have to face authors with excessive and unexpected charges, it would seem advisable to inform them before final action is taken.[14]

Adrian accepted the fatherly advice and promptly wrote to the authors reversing his decision to bill them for the mishap. He added, "I did not mean...to be critical either of your paper or of your work on the late lamented paper.... I can understand how you must feel after losing the results of hard work and good work, and I can also understand how it must feel to have a bill piled up on top of this."[15]

Adrian Albert was always one of the first to try out the latest technology. Soon after returning from Rio, we had our first television set.

Leonard Blumenthal saw our TV while attending one of Frieda's famous math parties and wrote afterwards, "Your set changed my ideas about television completely, and as soon as it is possible to get good reception here, I intend to buy one. As you observed, an ordinary broadcast seems flat after television."[16] The downstairs neighbors, the Boorstins, later recalled, "when we were young marrieds and you had the only TV set on the block, you were so nice to let our boys invade your living room."[17]

During the 1940s, Hyde Park still retained the character of a small town. Fifty-fifth Street was its bustling business center, providing both sustenance and entertainment. For the cost of a twenty-cent admission, the Frolic Theater served up a double feature accompanied by cartoons, newsreels, and a Superman serial every Saturday afternoon.

Often, I went grocery shopping with my mother on 55th. She made about half a dozen stops, selecting the best products in little specialty shops while ferreting out bargains at the chain groceries. For great baked goods, we stopped at Greenberg's Bakery. For mouth-watering bologna and caraway seed rye, we popped into the kosher deli. For canned goods, she patronized the chains—Kroger, A&P, and National.

Sometimes, we stopped at all three for their respective bargains of the week.

She overcame her original dearth of culinary skills and became an adequate cook. But her forte was pastry. She baked primarily to please her husband, who strongly encouraged her, unaware of the adverse effect that consuming her sugary products might be having on his health.

She poured her artistic expression into tiny tantalizing delicacies. At the same time, her bookkeeper's background showed in her productions, as she measured each ingredient with decided precision. The result was sweet and predictable perfection.

Her apple pies could melt the heart of Ebenezer Scrooge. There were cookies carefully crafted with the aid of cutters in sundry shapes and decorated with chocolate, nuts and colored sugar. Her pastries ranged from old favorites like brownies and mandelbrot to a seemingly unlimited assortment of gourmet delicacies that added an elegant, yet cost-effective, touch to the Alberts' math parties. Her crowning culinary achievement came as the result of a class in coffee cake cookery. The aroma of those twisted works of art, flavored with almond paste and vanilla extract, filled the Albert home with delicious anticipation of each masterpiece.

When we youngsters had money left over from our allowance, we would head over to 55th Street to buy a 5-cent ice cream cone from the fountain at McLaughlin's Pharmacy. Closer to home, we could buy five jelly beans for a penny at the tiny Mom and Pop grocery on Drexel Boulevard, just south of 56th Street.

Children felt safe then, even on the western edge of Hyde Park. In warm weather, we sometimes played in Washington Park near Cottage Grove or rolled down the hills on the Midway at 59th and Drexel. The campus was a good place to ride our bicycles. The 57th Street beach was a glorious place to while away the lazy days of summer.

At the time, no thought was given to restricting young girls' travel around the city. While in grade school, I began to travel downtown alone, commuting to Saturday afternoon classes at Chicago's Art Institute by way of the 55th Street bus and the Illinois Central train. On the train, I often chatted with a classmate headed for the same destination.

In June 1951, my Ray School classmates and I walked down the aisle to the cadence of Pomp and Circumstance. It was no surprise for me to see my father take the podium, since I had volunteered his services for the convocation address.

His talk was all about the short shrift being given to the subject of mathematics in America's schools. He complained,

> The battle against mathematics [as a required subject in high schools and colleges] was based on a plea that mathematics is too difficult to learn and should therefore not be required. This is false. The learning of mathematics is not unusually difficult and the fear of studying it is based on a tradition that some pupils cannot learn the subject. But there are really few bad students of mathematics. There are, instead, many bad teachers and bad curricula....The difficulty of learning mathematics is increased by the fact that in so many high schools this very difficult subject is considered to be teachable by those whose major subject is language, botany, or even physical education.[18]

He illustrated his point with this apocryphal story:

> The ignorance of some school administrators in this respect is no less than that of a country school board member who was visiting one of the schools under his jurisdiction. The teacher was nervous and anxious to please and she had succeeded in imparting her nervousness to the class. She asked one boy, "Who signed the Magna Carta?" The boy hesitantly replied, "I-I-I didn't do it!" The school board member, an old, tobacco-chewing backwoodsman, leaned forward and said, "Wait a minute, call that boy back. I don't like his manner. *I* believe he *did* do it."[19]

It was quite a relief afterwards, knowing that my father had done a fine job, and did not embarrass me.

My grade school friends and I would proceed to four full years of high school. But big brothers Alan and Roy had already chosen a different path. They both elected to follow the "Hutchins Plan."

In the early 1930s, as the newly appointed Chancellor of the University of Chicago, Robert Maynard Hutchins devised a curriculum fitting his concept that every college student "should have a solid knowledge of the foundations of the intellectual disciplines, should be able to use language and reason, and have some understanding of man and what connects man to man."[20]

By the time Alan came of age, Hutchins' concept had evolved. The Chancellor determined that college students needed not just two, but four full years of general liberal arts education before going on to specialize in any field. To accomplish this without substantially lengthening the number of years spent in school, he devised a program whereby precocious high school students could bypass one or two years of high school by enrolling in the College as "early entrants."

Hutchins expressed his philosophy of education in a compact tome entitled *The University of Utopia*. In it, he decried the effects of specialization in universities. He recommended that the work of the University of Utopia ought to begin with the junior year in high school and end with the sophomore year of college. "The object of the College of Utopia," he wrote, "is to see to it that everybody in it gets a liberal education."[21]

Alan enrolled at the University of Chicago in January 1946 after only two full years of high school. His college studies at Chicago were interrupted by the South American trip, but he did not seem to mind. He had a wonderful time in South America, and afterwards, went on to complete his bachelor's degree in June 1950.

The year I enrolled in high school, Roy was entering his second year in the college. I observed my brothers' lives as college students, but was in no hurry to join them. To me, high school was an adventure in itself and I had no desire to shortcut that adventure by rushing into college life.

Chapter 16

Albert Algebras

Weyl was right. Mathematicians never knew in advance whether the constructs of their imaginations would resonate with their fellow mathematicians.* Nor could pure mathematicians predict whether their inventions would ever play a role in the real world.

As a scientist, A.A. Albert ignored such concerns, allowing science to control the direction of his explorations. If science led him to a question he found challenging, he dove in and began to explore it. Like the detectives in his favorite mystery stories, he loved the thrill of the chase.

It appeared to be something of an afterthought when he produced a mere two-page paper in 1950 elaborating on the "algebra of quantum mechanics" he had discovered nearly two decades earlier.[1] The invention was a peculiar creature, since it resided in 27-dimensional space.[2] His brief paper revived the curious algebra that had languished since 1933.

In his 1950 mini-paper, Adrian simplified his earlier proof that the algebra he invented was an exception to Jordan's "r-number algebras." The new paper proved conclusively that the algebra he invented would not comply with the axioms applicable to those "special" Jordan algebras. The paper also expanded the potential utility of the "exceptional" Jordan algebra by deleting the restriction that it be finite-dimensional.

---

*   See Chapter 12 regarding Hermann Weyl.

Although the term was not widely used during his lifetime, "exceptional Jordan algebras" eventually came to be known as "Albert algebras," since he was recognized as their inventor. Both terms remain in use today.

After Adrian revived his discovery, several mathematicians, including his former doctoral student Richard Schafer, became interested in the apparent links between Adrian's discovery and other "exceptional phenomena" such as exceptional "Lie groups" and exceptional geometries.[3] In a series of papers produced in the 1950s, these mathematicians demonstrated the interconnections between Albert algebras, other topics in algebra, and other branches of mathematics as well.[4]

While these mathematicians had found new applications for Albert algebras, their research failed to bring about a better understanding of the algebras themselves.[5] Since Albert algebras had begun to generate a certain amount of interest, they seemed to cry out for further analysis. It fell to Adrian himself to unravel the nature of Albert algebras.

During 1956, he discovered a new type of Albert algebra—an exceptional Jordan "division algebra" which allowed not only addition, subtraction, and multiplication, but also division by non-zero elements to be carried out.[6] Jacobson confirmed that "[t]he first construction of exceptional Jordan division algebras is due to Albert."[7]

Adrian's finding contradicted the previous belief that there was only one type of Albert algebra, that is, a "reduced algebra" containing unusual idempotents (identity elements). His paper proved that Albert algebras could be partitioned into two distinct categories—"reduced Albert algebras" and "Albert division algebras."[8] The paper provided a basic cookbook of ingredients for constructing his new Albert division algebras, which he termed "cyclic Jordan algebras." Adrian's recipe sat on the shelf for seven years afterwards until he returned to tinker with it.

A German mathematician, Holger Petersson, later assessed the significance of the American professor's discovery of Albert division algebras: "The implications of this remarkable discovery can hardly be overestimated. In particular, it follows that exceptional simple Jordan algebras form a much wider class of mathematical objects than had been for a long time anticipated."[9]

Adrian further developed his Albert algebras in later papers. While polishing his paper on Albert division algebras, he coauthored a paper with Jacobson on "reduced exceptional Jordan algebras" (the term "Albert algebras" had not yet come into vogue). The duo gave a complete classification of "reduced exceptional simple Jordan algebras" over number fields.[10]

He collaborated with the algebraist, Lowell Paige, in another breakthrough known as the "Albert-Paige theorem."[11] Their theorem strengthened Adrian's original result by proving that the exceptional Jordan algebra is not the homomorphic image of any special Jordan algebra.

Reportedly, Paige first used the term "Albert algebras" shortly after his Chicago colleague's discovery of exceptional Jordan division algebras.[12] However, the term did not catch on until much later.

Few American mathematicians took an interest in the esoteric objects during the 1950s. Nathan Jacobson was a notable exception. The banter between the two men continued even when one of them left the country. Jacobson once wrote to Adrian from Paris questioning the existence of a solvable exceptional Jordan algebra. Adrian replied with a six-page paper titled "A Solvable Exceptional Jordan Algebra."[13]

During the mid-sixties, Adrian produced a follow-up to his paper on the theory of Albert division algebras (he continued to call them "cyclic Jordan algebras").[14] He advanced his "cyclic Jordan algebras" by presenting some new solutions and improving on the ingredients in his original recipe. Until then, his discovery had stirred little interest in the mathematical community outside of a few stalwarts.

He saw his idea catch hold toward the end of the sixties, when a new crop of mathematicians began to delve into Albert algebras. They recast the subject in terms more comprehensible to the next generation of scholars.[15] Jacobson pointed to the much simpler "rational" constructions given by the Belgian mathematician, Jacques Tits.[16]

Petersson and a colleague, Michel Racine, became interested in Albert algebras in 1975. They found the objects to be "highly complex mathematical structures enjoying a variety of remarkable properties that are both natural and mysterious."[17]

The Russian algebraist Efim Zelmanov studied Albert algebras and in 1979 proved Adrian had discovered the one class of exceptional simple Jordan algebras, and that there were no others.[18]

Beginning in the early 1990s, Albert algebras suddenly spawned a flurry of activity among mathematicians.[19] Petersson, who undertook an intensive study of Albert algebras, describes them as "an important algebraic structure that plays a significant role in various branches of mathematics and even physics."[20] Kevin McCrimmon, a former doctoral student of Jacobson, notes applications of Albert algebras to topics in genetics, probability, and statistics as well as various branches of algebra and geometry.[21]

John von Neumann made the observation that, most often, "there is a time lapse between a mathematical discovery and the moment it becomes useful...which can be anything from 30 to 100 years...the whole system seems to function without any direction, without any reference to usefulness, and without any desire to do things which are useful."[22]

It took close to 60 years for Albert algebras to attract the widespread attention of top researchers.[23] The peculiar algebra that came close to extinction rose from the ashes to begin a new life.

Chapter 17

Defending His Country

Having won the Second World War, the United States was not content to rest on its laurels. America's technological supremacy had played a large part in winning the war. In order to guarantee the future of democracy, America needed to maintain its technological edge.

A.A. Albert was already known to the Defense Department by virtue of his defense work during World War II. After the war, if there was a government agency doing any sort of defense work related to mathematics, he was almost certain to be involved in it.

America needed to expand the pool of young scientists after the lean years of the Depression and wartime attrition. The government decided to finance this effort by providing fellowships. Beginning in the late 1940s, Adrian helped design tests for screening potential postdoctoral fellowship recipients.[1] The work was a joint project of the National Research Council and the U.S. Atomic Energy Commission, a newly established agency designed to place development of the U.S. atomic energy program under civilian control.

He served the nation's defense in both advisory and "hands-on" capacities. The National Academy of Sciences appointed him to a three-year term on the Office of Naval Research (ONR) mathematics advisory committee in 1948 and reappointed him in subsequent years.[2] His old friend, Mina Rees, had signed on as head of ONR's Mathematics Branch two years earlier and played an important role in ONR, the first federal program in the United States to provide support for mathematical research after the Second World War.

For three months in the fall of 1951, Adrian took a leave of absence from Chicago to serve as on-site consultant to RAND (standing for "Research and Development"), a civilian think-tank. The atmosphere at RAND was electric. Some of the nation's top scientists brainstormed there on behalf of the U.S. Air Force. The top-secret project operated inside an unassuming office building in downtown Santa Monica, California.

RAND, originally a project of the War Department and the Office of Scientific Research and Development (OSRD), operated under the umbrella of the Douglas Aircraft Company when it started up in December 1945. In 1948, the Air Force approved spinning the project off into a private, nonprofit corporation, although contracts continued to come from the Air Force. Scientists like A.A. Albert who had consulted for OSRD during World War II felt a continuing obligation to support the defense effort. The fact that RAND was a civilian agency made it that much more palatable to academicians.

RAND's operations grew in size and importance after the Soviets detonated their own atomic bomb in August 1949. In June 1950, North Korea crossed the 38th parallel into South Korea and President Truman dispatched United States troops to the area. The Joint Chiefs of Staff interpreted the invasion by Russia's ally as a "Soviet ploy" to divert American military resources to the Pacific while Russia planned a full-scale attack on Europe.[3] Conspiracy theorists in America's military feared that the Soviets might find the Korean War an opportune time to employ their newfound atomic weaponry.

America's most important deterrent against a Soviet nuclear attack was the Air Force Strategic Air Command (SAC), the guardian of America's nuclear warheads. In 1951, the Air Force, concerned about the threat of Soviet attacks on Air Force strategic bombers scattered around the world, asked RAND for help in protecting SAC bombers and developing a counter-offensive plan.

RAND divided scientists into groups assigned to resolve specific practical problems. Adrian's analytical skills were put to use in determining the optimal strategy for bombing campaigns against multiple targets.[4]

Because the school year was starting, Frieda remained at home with the children. Adrian was unaccustomed to being separated from his wife for such a long period. He camped out on a cot at his brother's Los Angeles home and felt miserable. The separation prompted him to pour his heart out to her in a series of letters. "Every day seems like an age," he complained. "I want so much to be with you, to hear your voice, to see your sweet dear face."

He spent much of his free time with a large contingent of his wife's family who had relocated to the Los Angeles area. The separation was made a bit easier because television had swept the nation by 1951. He had no Frieda, but he did have weekend football and Sid Caesar.

Toward the end of his stay at RAND, his letters became more optimistic. In late November 1951, he wrote, "I have been...rejoicing that my period of exile is almost over."

Since he was busy working in Santa Monica that fall, he was obliged to decline two invitations. The first was a tea with Great Britain's Lord Halifax, who planned to visit the University of Chicago in October. The second was a reception at the Soviet Embassy in Washington, D.C. celebrating the thirty-fourth anniversary of the Great October Socialist Revolution.

Adrian returned to consult at RAND again during the summer of 1952. This time, he brought his family along. Coincidentally, while he toiled at the Air Force's RAND project, his son Alan served in the Air Force, stationed near the Air Force's B-29s based in Okinawa, where he worked on radar systems for the Strategic Air Command.

He continued to consult for RAND through 1955. During his consultancy, John von Neumann developed the JOHNNIAC, one of the first digital computers, at RAND. "Finite fields" formed the basis of that new technology. At the end of the RAND consultancy, Adrian published a text, sponsored by the Department of Defense, providing the modern theory of finite fields.[5] The book excluded certain information about the topic. He told the students in his course on finite fields, "There is a very interesting theorem I'd like to tell you about, but it's still classified."[6]

In 1952, his advisory functions for the National Academy of Sciences grew exponentially. He was appointed to a three-year term as Chairman of the National Research Council's Division of Mathematical Sciences.[7] The NRC is the principal operating agency for the National Academy of Sciences. The Academy was created by Congress during the Civil War as a private, nonprofit institution to advise the government in scientific and technical matters.[8]

Although the Academy functions essentially as an honorary society, since 1916, its NRC arm has played the substantive role of providing advice to the government. The NRC funneled out many of the research grants that went to universities and nonprofit institutions after World War II. The number and size of these grants increased at a prodigious pace in the wake of the war.

The NRC appointment required him to attend meetings in Washington, D.C. once every three weeks. As Chairman of the mathematics division, Adrian was obliged to attend all committee meetings as an ex-officio member. As a member of the NRC's governing board, he also attended board meetings five times a year.

This commitment, along with other advisory functions he performed during the 1950s, made a serious dent in the time available to pursue his first love—pure research. He constantly struggled to accommodate both priorities, refusing to acknowledge the impossibility of doing an unlimited number of things at once.

During the 1950s, the Chicago professor juggled several consultancies that occupied most of his summers. Besides RAND, he worked on applied mathematics projects operated by the Air Force Cambridge Research Center, Sandia Corporation, the U.S. Navy, and Project SCAMP. One of his first tasks was to advise the U.S. military in evaluating its latest technology for IFF—"Identification, Friend or Foe." The system was designed to transmit and receive identification codes to distinguish between enemies and friends in a tactical situation. The idea was that if they received the correct code back from a suspicious aircraft, the Americans would know it was one of their own and refrain from shooting it down. The government planned to spend millions equipping all U.S. military aircraft with the new equipment.

He examined the specifications for the technology and told the Generals, "Don't waste your money on it. The Russians would have no problem in foiling this system. Any mathematician worth his salt could do better than this." This advice prevented the military from squandering millions of taxpayer dollars and losing American lives in the process. Afterwards, the Defense Department found it useful to accommodate Professor Albert's wish to host summer research projects where mathematicians solved theoretical problems having practical applications for military security.[9]

Much of the applied mathematical research that he participated in was connected, directly or indirectly, with the National Security Agency. President Harry Truman established NSA under the Defense

Department in 1952 for the express purpose of unifying the work of disparate branches of the military relative to cryptography.[10] NSA, whose very existence was kept secret until 1957, was referred to by insiders as "No Such Agency."

NSA created a Scientific Advisory Board to bring the insights of academia to bear on the work of the new government agency. Adrian served on NSA's Advisory Board from 1953 to 1971.[11]

He also served on the Defense Department's General Sciences Panel beginning in 1953, and in 1954, was appointed to the 5-man steering group for the panel. The panel was an advisory group to the Assistant Secretary of Defense, who supervised research and development in the Defense Department.

During that period, the General Sciences Panel, with some urging from Professor Albert, declared all mathematics to be equally defense-related.[12] America was becoming awakened to the strategic value of supporting pure mathematics.

For three summers beginning in 1953, he participated in Project SCAMP (Southern California Applied Mathematics Project), a top-secret research project conducted on behalf of the Defense Department. Spending summers in California involved driving the family and a season's worth of belongings clear across the country. It was a delicate balancing act for Adrian. Frieda's idea of a trip out west was sightseeing at tourist spots like Yellowstone Park and Carlsbad Caverns. His idea of a fine trip was rousing the entire family at 2:00 a.m. in order to slice through the 1,700 miles separating Chicago and Los Angeles as fast as humanly possible. In the years before energy-saving speed limits, he revved up the engine, roaring down Route 66 at 80 m.p.h.

The ostensible purpose of SCAMP was to conduct research in "Numerical Analysis," but the project focused on cryptology. During 1959 and 1960, Adrian served as Chairman of SCAMP. He also organized and directed a top-secret project at SCAMP called "ALP," which researchers jokingly referred to as "Adrian's Little Project."[13]

I. N. Herstein later reported that many of the problems addressed at SCAMP and ALP touched on "field theory" and "permutation groups." According to Herstein, one of the dividends of the government-sponsored research was that many young mathematicians got their first taste of research while working at SCAMP.[14] Kaplansky recalled that, although researchers at SCAMP and ALP were told the ultimate objective was cryptanalysis, the problems they worked on were interesting problems in pure algebra such as factorization of polynomials and "groups given by generators and relations." Besides "group theory" and "field theory," researchers drew on "number theory" and statistics.[15]

Closer to home, Adrian played a role at the Argonne National Laboratory located some 25 miles southwest of Chicago. The University of Chicago served as principal contractor for the government-sponsored facility originally constructed during the Manhattan Project.

In 1942, while Fermi and his team worked on small scale graphite-uranium piles on Chicago's campus, planners selected the Argonne Forest Preserve location to host a full-sized experimental pile due to the need to move the potentially hazardous work away from densely populated areas. It was close enough to the "Met Lab," but not too close.

After World War II, many scientists believed that the facility would be closed. However, the Truman administration saw the need for a scientific research facility to maintain America's lead in the nuclear field. Argonne became the country's first national laboratory after President Truman signed the Atomic Energy Act of 1946.

The national laboratory's first mission was to develop nuclear reactors for both military needs and peaceful purposes. The program served to demonstrate America's superior military power to the world at a critical juncture, shortly after the Soviets had acquired their own nuclear program.[16]

As time went on, Argonne's research activities expanded. The Applied Mathematics Division, formed in 1956, provided support services such as mathematical analysis and programming for the Laboratory's computers. Adrian served on the institution's Applied Mathematics Advisory Committee and, from 1954 through 1961, he was Director of Defense for Research and Engineering at Argonne.

At the end of a Visiting Professorship at Yale University in 1957, he moved directly to his next job as Director of a "summer project" conducted at Bowdoin College in Maine. The summer project was one of the research efforts sponsored by the United States Air Force's Office of Scientific Research, Air Force Cambridge Research Center (AFCRC).

Adrian was well acquainted with Horst Feistel, who headed a "crypto group" at the Center.[17]

To the outside world, it seemed as though all of the people working in the project were algebraists conducting research in pure mathematics. In a sense they were, as some of the problems they researched that summer related to group theory.[18]

Although their research was in some sense theoretical, group theory had direct applications in cryptology. Feistel and his colleagues at the AFCRC prepared a long list of classified and unclassified problems related to cryptanalysis, including many "purely group-theoretic questions related to a type of cryptographic system then under investigation."[19]

The classified nature of the research project was disguised by calling it a "Summer Conference."* The relationship between group theory and cryptology was kept under wraps for years. In a letter written in the late 1960s, Adrian replied to a professor at a small eastern college who had asked about applications of group theory to cryptography. Adrian demurred, "I do not know of applications to cryptography of group theory...other than in material which is classified and where the applications are somewhat automatic."[20]

It was a gypsy-style existence—moving from place to place, going directly from a temporary abode in New Haven to another temporary home in Maine. The constant uprooting was a bit hectic for Frieda. She wrote to her children back in Chicago, "I seem to spend a good part of my life packing and waiting for the Express men."[21]

A.A. Albert was elected Chairman of the National Science Foundation's mathematics section in 1957, a time when the country was filled with a sense of angst over *Sputnik*. The agency had a heightened mandate to increase the number and quality of scientists in light of the Soviets' demonstration that it could place a satellite into orbit. To Washington, the Soviets' ability to launch *Sputnik* implied that they could also send hydrogen bombs to the United States atop intercontinental ballistic missiles. One of Adrian's roles at NSF was serving on a planning committee to set budgetary needs for mathematics.

---

*   More about the "Summer Conference" in Chapter 20.

Unlike the National Research Council, with its 40-year history, the National Science Foundation was then a mere seven years old, having been created by Congress in 1950 to promote the progress of science in the wake of World War II. His new position added one more administrative job to Adrian's plate, since he had held the post as Chairman of the National Research Council's Division of Mathematical Sciences for half a decade. Students reported that he appeared somewhat distracted in class during the 1950s.

The NSF's mission was to support basic research in the sciences and to evaluate the research programs undertaken by government agencies.[22] Unlike the National Research Council, NSF was a government agency with a board appointed by the U.S. President.

The job as NSF mathematics chair came on the heels of Adrian's work in chairing an NSF Survey on Training and Research Potential in Mathematics between 1953 and 1957. Known to some as "the Albert survey," the project sought to assess the nation's training needs and provide a baseline of information for future funding by the NSF.

Fortunately for Adrian, a capable mathematician named John Green served as Principal Investigator for the project, and took the brunt of tabulating questionnaires from his shoulders.[23] Adrian was not shy about letting Green and others know how much he appreciated Green's work, since the success of the project depended on it.[24]

The Institute for Defense Analyses was founded in 1956 with a $500,000 grant from the Ford Foundation to aid the National Security Agency in securing top academic scientists to work on the national defense.[25]

Although IDA was not directly affiliated with any university, representatives of a dozen universities were placed on IDA's board in order to lend a certain amount of credibility to the new venture. Adrian began consulting for the Institute for Defense Analyses in 1959 and later became the University of Chicago's representative on IDA's Board of Trustees.[26]

In 1959, IDA received almost $2 million from the government to construct and develop a laboratory for basic research into "defense-related communications."[27] The term tended to obfuscate the fact that research on cryptology would be conducted at the lab. The IDA lab would be headquartered on the campus of Princeton University where Adrian had spent his postdoctoral year.

In a sense, the IDA lab was an outgrowth of the summer research projects Adrian had organized. It centralized all previous cryptology projects in one agency. The NSA selected him as the lab's first director to get the project up and running.

During 1961–62, he took a leave of absence from the University of Chicago to spend the academic year in Princeton as Director of IDA's new Communications Research Division. His presence contributed more than the substantive research he produced that year; it lent a legitimacy to the new venture that persuaded other academics to commit to working at the institution.[28]

The focus of his research at IDA was on codebreaking. The classified research he conducted there concerned very specific problems. In working out the solutions, he reduced the problems to abstract mathematical concepts, thereby allowing a sanitized version of some of the research to be declassified afterwards.[29]

The townspeople in Princeton sensed that some sort of top-secret work was being done at the IDA facility, but could only guess at the specifics. Waitresses at the local lunch counter would occasionally interject queries such as "Hey, do you guys have your fingers on the A-bomb button?" The researchers would simply smile and continue to eat their lunch.

To aid the nation's air defense and space race, Adrian consulted with Defense Department contractors during the 1960s.[30] At MITRE Corporation, he used his knowledge of information systems technology on behalf of MITRE's electronically based air defense systems. At the Jet Propulsion Laboratory of the National Aeronautic and Space Administration, Adrian assisted the U. S. government in its competition with the Soviets to put the first spacecraft in orbit around the earth.

Mathematics provided the underpinnings for telemetry, the technology that enabled spacecraft to employ data encoding used in data reception and tracking of spacecraft. Specifically, he was retained by JPL to provide "information regarding the properties of finite projective planes and related algebraic structures derived from incidence matrices for ranging and telemetry codes."[31] When NASA's Mariner 2 coasted noiselessly past Venus on December 14, 1962, American mathematicians such as Adrian Albert contributed to NASA's success.

In recognition of his efforts on behalf of the national defense, the retiring Director of the National Security Agency wrote in 1969 expressing his "tremendous appreciation" for Adrian's contributions as a member of the NSA.[32]

Chapter 18

The University's Solution to Urban Decay

After the war, the United States government attempted to help our conquering heroes make up for time lost in the service of their country. Congress enacted the GI bill, and tens of thousands of veterans enrolled in college. The Veterans Administration underwrote no-money-down home loans to jump-start a normal life for veterans and their families.

Coming on after the lean years of wartime rationing and housing shortages, the federal infusion of funds fueled a building boom. To meet the sudden need for housing, new subdivisions sprang up. In the Chicago area, developers built new homes in such inner ring suburbs as Evanston and Morton Grove.

Meanwhile, decay set into Chicago's inner city. Its older housing stock and overcrowded conditions did not appeal to the veterans and their young families.

Hyde Park felt the change as families fled the city in droves for greener pastures. You could see it in the declining number of small shops and groceries on 55th Street. Like the families, the shop-owners fled. Less savory businesses now stood in their place. Researchers counted 23 taverns in one two-block stretch of 55th Street.[1]

Urban decay began to gnaw at the very foundation of that venerable institution of higher learning, the University of Chicago. Crime on the streets was driving both faculty and students to seek out universities in safer cities. The "brain drain" threatened the University's very existence.[2]

The Albert family, too, noticed the change. Fifty-seventh and Drexel lost the ambiance of a quiet, middle-class neighborhood. Buildings just west of ours were carved into crowded tenements. The buildings' occupants engaged in loud, angry exchanges in the middle of the night. The Alberts decided that this was no longer the best neighborhood for raising their children and began to search for another place to live.

Residents of Hyde Park and the neighboring community of Kenwood became concerned about rising crime rates, mass emigration, and physical deterioration.[3] Several grass roots organizations sprang up to address the problem. In 1949, the community banded together to form the Hyde Park-Kenwood Community Conference and prepared to fight blight.

Hyde Park held a trump card that other decaying neighborhoods in the city lacked—the University, with its $200 million investment in physical plant, highly resourceful faculty living nearby, and influence, both locally and nationally. Sparked by a series of crimes in the area involving University faculty, the University held a public meeting in March 1952 attended by some 2,000 Hyde Park and Kenwood residents. The South East Chicago Commission (SECC) was formed as a result of the meeting.

The University contributed $15,000 to the Commission's first annual budget and pledged at least $10,000 for each of four succeeding years. In the fall of that year, the University appointed Julian Levi, brother of then law professor Edward Levi, as Executive Director of the Commission.

The Commission and its ally, the Hyde Park-Kenwood Community Conference, attacked the problem on several fronts. These grass roots organizations, along with the University's Chancellor Lawrence Kimpton, went hat in hand to the Chicago Land Clearance Commission, asking it to survey the community for blight. The Land Clearance Commission was a municipal corporation created in 1947 by a state law known as the Blighted Areas Redevelopment Act. That law gave the agency sweeping powers, including the powers to acquire property by eminent domain, borrow money, apply for federal funds, and sell the acquired property in accordance with a redevelopment plan.

Another group was simultaneously instrumental. Early in 1952, the Metropolitan Housing and Planning Council, a civic association concerned with planning in the Chicago metropolitan area, set up a Conservation Committee to study how declining neighborhoods could be shored up through government and private planning. At the time, more than one-third of the Council's board members either lived in Hyde Park-Kenwood or were graduates of the University of Chicago. They produced a report recommending Hyde Park-Kenwood as a pilot project.

In July 1952, the Chicago City Council passed a resolution calling for the establishment of an Interim Commission on Neighborhoods. This Commission, headed by James Downs, Jr., worked closely with the Metropolitan Housing and Planning Council.

Downs' Commission delivered a report calling for the creation of a permanent Community Conservation Board to undertake projects to renew Chicago's neighborhoods. This recommendation became a reality in 1953, shortly after the state legislature passed enabling legislation known as the Urban Community Conservation Act. However, since, with typical bureaucratic lethargy, the Board did not become fully operational until much later, the University and the SECC took the reins and began preliminary planning for Hyde Park's urban renewal. Their work was sponsored by a $100,000 grant to the University from the Marshall Field Foundation.

The University's receipt of this grant escalated matters one step further. Jack Meltzer became director of the SECC's Planning Unit with offices on the University's Chicago campus. This ultimately led to the City of Chicago's contracting with the University to produce an Urban Renewal Plan. Which, in turn, led to the Chicago Land Clearance Commission's designation of two areas within Hyde Park as "slum and blighted area redevelopment projects" as defined by the state's 1947 Blighted Areas Redevelopment Act. The Chicago City Council and the Illinois State Housing Board approved the Commission's action in July and August 1954.

These actions were well and good, but did not suffice to bring the University's plans to fruition. Federal funds were the fuel needed to pump the engine of urban renewal. At first blush, Messrs. Meltzer and Levi looked to the federal Housing Act of 1949 for help, as that law authorized federal assistance for slum clearance. However, the Act was limited to clearance and redevelopment of slums, and would not have extended to a "conservation" area such as Hyde Park. An amendment would be needed to expand the scope of the law.

Help finally came in August 1954, when the United States Congress amended Title I of the 1949 Housing Act by authorizing federal assistance to prevent slums and urban blight through rehabilitation of deteriorating neighborhoods. Under the expanded law, the administrator of the federal Housing and Home Finance Agency was authorized to provide two-thirds of the funding for clearing and preparing redevelopment sites, with the remainder to be provided at the local level.

Simultaneously with the amendment, Congress passed the Housing Act of 1954. The new law implemented Title I by enabling the Federal Housing Administration (FHA) (then a constituent agency of the Housing and Home Finance Agency) to insure urban developments.[4]

The legislation created a huge boon for real estate developers. Under the new law, the Federal Housing Administration (FHA) was authorized to insure urban developments. It meant that developers need pay only 3 to 5% of redevelopment costs. Lenders would supply the rest, guaranteed by the full faith and credit of the U.S. Government. Interest rates on these loans were capped at 4.25%.

Meltzer and Levi drew up a detailed redevelopment plan and submitted it to the FHA. Their plan called for buying about 32 acres of land in Hyde Park at a cost of about $6 million. The land would be used to build high rises, townhouses, a shopping center, recreational area, and parking.

Their plan sailed through the City Council and the Illinois State Housing Board. In March 1955, the federal Housing and Home Finance Agency signed, sealed, and delivered the contract. During the years that followed, Hyde Park would begin its transformation.

Chapter 19

East of the Tracks

The Albert family could not wait for redevelopment. Soon after my graduation from Ray School, we said goodbye to our old off-campus apartment building. Frieda had no regrets. We were moving to a better area—East Hyde Park. Due to the added distance from the University, the Alberts became a two-car family. Adrian's new mode of transportation was a tiny Opel with a stick shift.

All of the land lying between the lakefront and the Illinois Central tracks constituted East Hyde Park. Our new abode was a second-story apartment in an Art Nouveau graystone and brick six-flat. That 1907 beauty boasted soaring Gothic parapets, stained glass windows, a tile roof, and an ornate stone entrance guarded by a pair of gargoyles. Architects Fromman and Jebson designed the building to blend in with the luxurious single-family residences lining Hyde Park Boulevard at the turn of the century.[1]

The juxtaposition of tenants in the vintage structure seemed strangely symbolic. Our apartment rested directly beneath the third floor residence of Adrian's boss, Walter Bartky, Dean of the University's Physical Sciences Division. Adrian's Mathematics Department colleague and billiards buddy, Harry Everett, lived in the apartment across from ours.

Mahogany beamed ceilings graced the living and dining rooms of our oversized apartment. Its bowling lane-like hallway spanned half a city block.

Paneled mahogany pocket doors divided the enormous living room from the library, where Adrian would sit at one of the desks, puzzling over mathematical theorems. His nest of pipes sat atop the right-angled bookcase in the corner of the library. Frieda had encouraged him to take up pipesmoking because it looked professorial.

The remainder of our books rested on the built-in bookshelves flanking the living room fireplace. At Christmas time, the shelves glittered with dozens of colorful cards, many from mathematical families in faraway places. I often curled up in that alcove, admiring the festive cards and family photos.

The living room flowed out onto a stone balcony—a picture perfect setting for summertime parties.

Glass-paneled cabinets in the massive dining room and adjoining butler's pantry displayed stemware and dishes. The butler's pantry was where Frieda kept the hard drinks she served at math parties. She was terribly allergic to alcohol, and could not even tolerate it in her food. Adrian despised strong flavors, especially those in mint, vinegar, and any form of alcohol. His taste in food ran to bland meat and potatoes, seasoned exclusively with salt. He urged her to stop serving hard drinks at their parties altogether, but she was too savvy a hostess to give in on that point.

The apartment boasted two gas-burning fireplaces. They did not interest Frieda—she placed furniture of front of both of them. A phonograph cabinet blocking the fireplace set the tempo of the living room. Adrian's favorite vinyls stood quietly inside its blond wood doors. An eclectic mix of music and mayhem, his collection ranged from comic routines by the likes of Tom Lehrer, Shelly Berman, and the Nichols-May duo to a sprinkling of classical pieces. Contemporary American folk music dominated the collection. After dinner, the room might reverberate with the ballads of Harry Belafonte, Judy Collins, and Joan Baez or harmonies by groups such as the New Christy Minstrels, Kingston Trio, Weavers, and the Limeliters.

I soon made friends with two teenaged girls living opposite the Bartkys. Their father was vice-president of a Fortune 500 company. They attended a private high school, St. Xavier, and belonged to the South Shore Country Club where their family played tennis during the summer. It was an exclusive club—no minorities allowed, including mine.

My friends' family had a live-in maid in their maids' quarters—a bedroom and bath off the kitchen. No one slept in our maid's quarters, since the Alberts' household help came only twice a week. In accord with society's emerging egalitarian sensitivities, we eschewed the term "maids," referring to our employees as "cleaning ladies."

Mrs. K. began working for the Alberts shortly after arriving from Mississippi in 1951 and remained in the Alberts' employ for the next fifty years. She revealed that Frieda was the first white woman to lunch with her at the same table. Mrs. K. later described Adrian to his granddaughter. "They don't come any better," she said.[2]

It was only a two-block walk from home to the 55th Street Promontory—a small peninsula of parkland jutting out into Lake Michigan. We knew it as "the Point"—a glorious place for a young girl entering her teen-age years to socialize. In the summers, the Point was strewn with beach blankets full of sun worshipers. Many were chattering Hyde Park High Schoolers whiling away the lazy days. It was our hang-out. Everyone you'd want to see was there.

Even my brother Roy stopped by our blanket to show off his new "Charles Atlas" physique. At five foot six, Roy overcompensated for the fact that he failed to grow one year due to his diabetes. Spending the summer on a construction job lifting oil barrels proved he was not a 90-pound weakling. My friends thought he was gorgeous.

Anyone who has lived in Hyde Park has a favorite story about the Point. Kaplansky recalls, "The swimming is great in the summer (although not exactly legal). I have a clear recollection of getting the idea for Lemma 5.... It happened while sunbathing in between two swims."[3]

Math parties at the Alberts' Hyde Park Boulevard apartment could be rollicking affairs. By then, Kaplansky had found and married his lifelong mate, the vivacious Rachelle ("Chellie"). Kap would sit at the piano playing show tunes and Tom Lehrer songs, surrounded by his wife and other Department members like Professors Paul Halmos and Shiing-Shen Chern with their wives.

Chellie Kaplansky, a warm, charming woman with flowing black tresses, shared her husband's musical bent. One year, she starred in the "Revels," the faculty club's annual spoof. For her role, she donned a Hawaiian costume, sang, and performed a hula dance while her husband whistled and shouted from the audience, "Go, Chellie!"

Adrian and Frieda maintained close ties with many of their old school chums, some going back to high school or even grade school. In 1946,

Adrian was voted President of the Marshall High School Alumni Association.

Some of their friends were affiliated with the University; others were not. The couple had one group of friends they referred to as the "Canasta gang." In a whimsical poem, a member of the "Canasta gang" decried the unfairness of competing with a mathematician at cards:

> The tale we tell is weird indeed,
> How Adrian fell, a victim of greed.
> For anyone else, it wouldn't be so bad,
> For a guy in the numbers racket, it is sad!
> Whenever there's a Canasta clash
> He sits there winning all the cash.
> Because our algebraic wizard
> Knows the rules from a to izzard.

At the end of the poem, the author rejoices at finally prevailing over her formidable adversary and forcing him to "eat humble π."[4]

Hyde Park offered plenty of opportunities for youngsters to socialize, or become socialized, depending upon one's perspective. As a ten-year old, I joined the Girl Scout troop at the Hyde Park Baptist Church. As a teenager, I enjoyed the socials at Chicago's beautiful Sinai Temple overlooking Lake Michigan and joined a girl's club at the Hyde Park YMCA.

Hyde Park High held a reputation as a top-notch public school. Year after year, Hyde Park students landed citywide awards in mathematics and Latin. The school had its share of optional activities that enriched the curriculum. I enrolled in Creative Writing and joined the A Capella Choir.

Jackson Park lay right across the street from Hyde Park High and the school availed itself of this amenity. The girls' gym class took many a hike through the park's Wooded Island—a nature sanctuary constructed for the 1893 World's Fair.

As graduation approached, I teamed up with two friends to pen a nostalgic farewell to the school where I had spent some of my happiest years. The three of us sang our "Class Song" at Commencement.

While idyllic for me, the early fifties were troubled times for the nation. America joined the war between North and South Korea. Big brother Alan enlisted in the U.S. Air Force and shipped out to the far east. For a time, he was stationed in Okinawa. Later, he was reassigned to South Korea, where he worked on the Air Force's radar systems. After four years in the service, Alan was released on September 15, 1954, a day he referred to as "Golden Day."

PART FIVE

Elder

Statesman

Chapter 20

Years of Triumph and Tragedy

Gilbert Bliss retired as Chairman of Chicago's Mathematics Department in 1941. Adrian Albert served as Acting Chair on and off between 1943 and 1957 while others were named to head the Department.* Like the proverbial bridesmaid, he had yet to receive the ring.

As Acting Chair he attempted to spur some changes in a Department that had weakened during the war. Believing it suffered from poor leadership after Bliss' retirement because of the tendency to appoint professors from the ranks of former U. of C. grad students, he went to Dean Bartky to urge that "something had to be done about nepotism in the Department".[1] They discussed several outstanding mathematicians including Marshall Stone and Paul Halmos.

In the fall of 1946, the Acting Chair engineered an assistant professorship for Paul Halmos, whose 1942 book, *Finite Dimensional Vector Spaces*, had received rave reviews for mathematical exposition.

Marshall Stone was not just a fine mathematician. He was also the son of United States Supreme Court Chief Justice Harlan Fiske Stone. Chancellor Hutchins attempted to recruit Stone by offering him a distinguished service professorship. At the time, Stone was ensconced as head of the Mathematics Department at Harvard. He held out until Hutchins agreed to grant him the chairmanship at Chicago.

Adrian wrote to his confidante, Solomon Lefschetz, about the appointment. "I feel that Stone will be able to attract both students of

---

* A.A. Albert served as acting departmental chairman during the following quarters: Winter 1943, Autumn 1946, Autumn 1949, Spring 1950, Winter 1952, Winter 1953, Winter 1954, Autumn 1955, Winter Spring and Summer 1956, and Autumn 1957.

185

the type we want and good faculty men. I am personally very happy that Stone is coming, if a bit surprised."[2]

Although Stone arrived from Harvard, he came with an interesting credential for the post at Chicago. He was in some sense a "mathematical grandson" of Chicago's first chairman, since Stone had studied for his Harvard doctorate under George D. Birkhoff, a first generation student of E.H. Moore.

Stone accepted the appointment because he saw the opportunity to place his stamp on a department that had been depleted by the retirements and resignations of five professors. He hoped to fill the vacancies with a new set of first-rate mathematicians and to modernize the curriculum.

The easterner had his regrets after accepting. He complained about Chicago's "hand-to-mouth" budgeting practices.[3] Stone felt obliged to threaten the University's Vice-President that he would resign if the appointment of S. S. Chern was not approved. He succeeded in revitalizing the Department with a series of appointments that included Chern, Antoni Zygmund, André Weil, Saunders Mac Lane, and two younger men, Irving Segal and Edwin Spanier.

Acquiring such a panoply of superstars was no mean achievement. The cantankerous Weil was the dominant member of the Bourbaki group of French mathematicians who successfully promoted rigor and abstraction in mathematics. Zygmund was on the road to becoming the central figure in the American school of classical Fourier analysis. Chern would become a world leader in differential geometry. Stone's friend and former Harvard colleague, Mac Lane, had earned high praise for his work in coauthoring a well-written text on modern algebra with Garrett Birkhoff, the son of G.D. Birkhoff. Together with A.A. Albert, Halmos and Kaplansky, the new hires formed the core of the Stone department at Chicago.

Felix Browder, who later ascended to the chairmanship of Chicago's Mathematics Department, gave the "Stone age" the highest accolades, declaring that Stone "transformed a department of dwindling prestige and vitality once more into the strongest mathematics department in the U.S. (and at that point probably in the world)."[4] Stone remained as Chairman until 1952, when, weary of the struggle with the administration for new resources, he stepped down.[5]

Adrian would wait a while longer for his turn at the helm. In 1952, the new Chancellor of the University, Lawrence Kimpton, appointed

Saunders Mac Lane as the next Chairman. Mac Lane stayed on as Chairman until 1958.

In the meantime, Adrian had plenty of things to keep him occupied. Lest he have a spare moment, he was besieged by requests from the editors of *Encyclopaedia Britannica* to work on write-ups for frequently updated editions.

In June 1952, he learned of his appointment as Chairman of the National Research Council's Division of Mathematics.[6] The assignment entailed serving on a variety of committees.

It seemed as though he had as many off-campus as on-campus activities. He logged 40,000 miles in air travel during 1953. He chaired or was active in about nine separate groups nationally while simultaneously consulting for the National Science Foundation, the Air Force Cambridge Research Center, and the National Security Agency. Under the auspices of the National Science Foundation, he chaired a committee responsible for a summer research project on Lie algebras held at Colby College in Maine during the summer of 1953.[7] Meanwhile, he also served on a number of boards and committees at the University.[8] Frieda complained that he did not know how to say "No."

At a meeting of the National Research Council, he attempted to decline the chairmanship of a survey on mathematics training in U.S. universities. He reported to the administration at Chicago, "I tried to get them to select someone else...but they insisted that I serve. I was also told that the National Science Foundation would not support the affair unless I agreed to take full charge. The survey is a challenge and should have important consequences for the welfare of mathematics. It will take a fair amount of time, but I will have the services of a full-time investigator and this should relieve me of a great deal of drudgery connected with the task."[9] As described earlier, that investigator, John Green, helped bring the project to a successful conclusion.

On campus, he continued to conduct his research, teach, and supervise doctoral students. His lectures were more workmanlike than slick or flashy.[10] J. Marshall Osborn recalled that, during the mid-fifties, "he was not the best teacher, but he did a good job. I got used to his style. Sometimes, he made mistakes in class, like when he did some calculations that were different from the calculations in one of his papers. It turned out, the paper was right after all."[11]

His teaching style was "detailed and methodical. If you had class notes from one of his courses, you had the whole course in front of you. It was all there."[12]

Adrian's lectures had their lighter moments. He had a unique way of explaining those subsets of mathematical rings known as "ideals." He would write the description of "left ideals" with his left hand and proceed to describe "right ideals" using his right hand. Then, he wrote the description of "two-sided ideals" using both hands simultaneously.[13]

The ambidextrous teacher's habit of writing with both hands drove some of his students to distraction. He often erased the formulas he had just written with one hand while writing with his other hand, leaving his students breathless. Murray Gerstenhaber, who received his doctorate under Adrian's supervision in 1951, paraphrased the French mathematician Hermite for the proposition that the reason his generation had so many good mathematicians was because they had lousy teachers.

One of the classes A.A. Albert taught at Chicago was Advanced Matrix Theory. He introduced the class to a theorem that sounded like a very simple mathematical statement about square matrices. However, the deceptively simple theorem was very difficult to prove. Boiled down to its essentials, the theorem stated that every matrix of trace† zero is a commutator.‡[14]

To aficionados of matrix theory, it was a beautiful theorem.[15] Emmy Noether's student, the Japanese mathematician K. Shoda, became quite interested in this matrix theory problem during the 1930s. Shoda solved the theorem in 1937 for fields of characteristic zero, but Shoda's proof was invalid for other fields.

Afterwards, almost 20 years passed before anyone could crack the problem. Although it had yet to find any practical applications, it was a puzzle begging for an answer.

Adrian began working on a solution that would hold for all fields, but had stopped short of finding the complete solution. Then it occurred to him that here was a nice academic problem ideally suited to a graduate level class in matrix theory. He wrote the problem on the blackboard and, like the mythical Professor Lambeau in the 1990s movie "Good Will Hunting," challenged the students to complete his proof.

---

† The "trace" of a square matrix is the sum of the elements in the main diagonal.
‡ A "commutator" is a particular type of element that we need not define here.

The ending to the story was not at all Hollywoodesque. The difficult problem was not solved by a hapless school janitor, nor did a solution come effortlessly overnight.

After allowing the problem to percolate at the back of his mind over several months, a second-year graduate student named Benjamin Muckenhoupt brought in a typewritten solution and tentatively handed it to his professor. The student, son of a Boston-area mathematician, had come to Chicago with a bachelor's degree from Harvard. Professor Albert took the proof from Muckenhoupt and examined it with a skeptical eye.

Adrian went over the solution with a fine tooth comb. Several days later, he called Muckenhoupt into his office to render his verdict. Muckenhoupt half expected the worst—a humiliating exposition of fatal flaws in his proof. He listened intently as his professor informed him that the proof for matrices of trace zero contained no trace of error whatsoever. The proud professor demonstrated his delight with Muckenhoupt's accomplishment by publishing the results of their collaboration in a joint paper.[16] The paper was a coup for Muckenhoupt and helped to launch his career as a researcher of pure mathematics.

A.A. Albert was not one to let bureaucracy stand in the way when there was a greater purpose to be served. As Acting Chair in the early 1950s, he learned that Paul Cohen, a young Putnam Competition winner, was languishing at a local New York college due to lack of funds. He wrote a letter urging Cohen to apply for graduate school at Chicago, mentioning that "finishing college would not be necessary."[17] Cohen flourished under the influence of Chicago's "Stone age" mathematicians and gradually blossomed into one of America's most accomplished mathematicians, winning top honors for his work on the foundations of set theory.

The Albert family celebrated Adrian's fiftieth birthday in 1955. Middle child Roy penned a few verses in honor of the occasion: A sample of a symbol in a solipsistic vein;/ A ripe, romantic ripple in a quite archaic train—/Perceive it on all levels. Fifty years no bitter pill./May your years fit as a mantle as you travel through life's mill./May your days be lit with wisdom and your nights with slumber sound./Health and power remain unflagging, the next fifty roll around./Esteem, respect, attention, all are tributes of the land./But although these sometimes lacking, you are loved close by at hand.

Most of the tributes bestowed upon Professor Albert came after years of volunteer work. The American Association of University Professors elected him President of its Chicago Chapter for the term commencing in January 1955. He informed the administration at the University of Chicago, "I continue to work with them as Editor of a proposed new series of texts on mathematics."[18]

In 1956, the American Mathematical Society voted him Vice-President-Elect. His 1957–58 term as AMS Vice-President followed years of service as chairman of various AMS committees and editor of AMS publications.

I decided not to follow in my brothers' footsteps. Instead of matriculating at Chicago, I opted to major in journalism at the University of Wisconsin. Madison was a lovely place, with its quaint lake and rolling hills. But I soon came to feel like a fish out of water. I was not cut out for a "big ten" school and begged my parents to allow me to transfer to Chicago.

By the time I entered the University of Chicago in January 1956, the "Hutchins plan" had gone the way of the dinosaurs. It was a grand Utopian experiment that, by most accounts, ended badly.

The "early entrant" program proved to be a disaster for many students, some as young as 14. It was their first venture away from home. Some were thrown in with WW II veterans many years their senior. The confused youngsters could not measure up to their more worldly-wise classmates, and many had trouble adjusting socially.

Under the Hutchins plan, students were not required to attend class. Their grades depended upon one make-or-break "Comprehensive Examination" in each class administered at the end of the school year. It was taken for granted that these brilliant students possessed enough self-discipline to manage their own time effectively.

Perhaps Chancellor Hutchins and his mentor, philosopher Mortimer Adler, had envisioned professors roaming the ivy-covered halls like Socrates, debating great ideas with students. In actuality, a significant contingent of the student body avoided attending classes altogether, preferring to pass their time studying the sciences of bridge and poker in the Reynolds Club's student lounge or at one of the nearby frat houses. When "comp time" approached, these students held all-night vigils, fortified with No-doz, cramming for the all-important exams.

Some vestiges of the Hutchins plan remained when I entered the University. All students were subjected to a series of placement exams for the liberal arts curriculum. Theoretically, a student could place out of the entire 14-course curriculum and earn a liberal arts diploma without ever setting foot on campus.

Now that I had left Madison, I longed to take part in campus life at Chicago and persuaded my parents to let me live in the dorms. A thoughtful University administrator assigned me to a comfortable room in beautiful Green Hall with one, quite compatible, roommate. I was thrilled, since in Madison I was among nine living in an off-campus attic. At Wisconsin, out-of-staters were not admitted to state-owned dormitories.

Green was one of several turn-of-the-century women's dormitories strung together into a single long building facing University Avenue. Residents congregated in shared lounges and a dining hall. Male students were allowed in the main floor lounges. "Hootenannies" often enlivened Friday nights. A popular guitar-playing student from one of the men's dorms led the folk singing, surrounded by a bevy of coeds.

University men had the option of livng in the dorms or joining one of the national fraternities that did double-duty as dormitories. The fraternities held frequent weekend parties. At one of the more interesting frat parties, a tall, handsome fraternity member brought out his pet boa constrictor and dangled it around his neck all evening.

Sororities were strictly forbidden on campus. It was whispered that sororities had been banned due to a tragic incident involving a female student who committed suicide after being blackballed by a sorority. According to the rumor, that student was the daughter of a major donor to the University. Women's clubs on campus served the University's female population, although they owned no houses and thus were not on a par with the fraternities. The arrangement was fine with me. As far as I was concerned, dormitory life at Chicago was as good as it gets.

While I had loved the ambiance of the campus as a child, I relished it still more as a college student. Being a part of the college made roaming around the Quadrangles that much more enjoyable.

Without a doubt, my favorite spot on campus was Bond Chapel, that tiny, tranquil refuge in the midst of the bustling Quadrangles.

Whenever I passed by, its heavy wooden door stood wide-open, inviting visitors to step inside. I would often wander into that unoccupied space and reflect on the beatitudes adorning the chapel's walls.

Chicago had a very open campus. One could stroll into any building, no questions asked. At first, I did all of my studying at Harper Library, which housed the main reading room at the University.

That ivy-covered English Gothic masterpiece, embellished by soaring towers and enchanting gargoyles, harkened back to the architecture of chapels at Oxford and Cambridge, England. The vaulted ceiling of Harper's massive reading room evoked images of decades of scholars laboring beneath its august arches. William Rainey Harper died in 1906, six years before completion of the magnificent edifice dedicated to his memory.

Later on, I varied my study habits, searching out more intimate reading environments. I explored the cloister between the botany and zoology buildings until I discovered a small, out-of-the-way reading room. I ventured into the hushed library at the Divinity School's Swift Hall. Eventually, I found the Law School's stately reading room where only a hundred or so of the 400 seats were filled with aspiring legal scholars.

One could arrive at the Law School's reading room by climbing two flights of stone steps from the school's lobby. Or one could cross over the connecting bridge from Harper Library's third floor reading room.

Few women traversed that passage. As far as I knew, only two women attended classes at the Law School. One lone woman served on its faculty, the famed Soia Mentschikoff, known for her work on the Uniform Commercial Code. Several college coeds like myself chose the reading room for their studies. Our presence was welcomed in that nearly all-male bastion.

The basement of the Law School housed a small lounge equipped with overstuffed chairs and coin-operated beverage machines. On several occasions, I noticed a tall, powerfully built man wearing suspenders and a plaid shirt wander in, stick some coins into the coffee machine, and exit with his cup of java. Given his physique and mode of dress, I assumed he was the janitor.

Little did I know that, years later, I would see the man on television. That unassuming gentleman was Professor Nicholas deBelleville Katzenbach, who became Deputy U.S. Attorney General in the

Kennedy administration and later, advanced to Attorney General for the Johnson Administration. In the midst of a reign of terror waged by the Ku Klux Klan throughout the southern states, Katzenbach would face down Governor George Wallace on the steps of the University of Alabama to personally supervise safe passage for two young black students.

Faculty life at Chicago's Mathematics Department resembled a game of musical chairs. Adrian filled in as Acting Chairman for a full year commencing with the fall quarter of 1955 while Chairman Mac Lane took a leave of absence.

The following academic year (1956–57), Adrian would leave for a Visiting Professorship at Yale. On the eve of his departure, Warren Johnson, Dean of Chicago's Physical Sciences Division, sent a note. Johnson thanked him for "remarkable leadership" in solving a number of problems in the Mathematics Department during his acting chairmanship and closed with the sentiment, "We are all looking forward to having you back again a year hence." The letter had the ring of a Dean concerned about the possibility of losing one of his own to an ivy league school.

The Alberts' elder son Alan and his wife had been renting a small apartment in Hyde Park since their marriage in June 1955. They gladly moved in to house-sit the Alberts' huge apartment while the parents were away at Yale. It was a fitting time for the Alberts' son to shelter under their roof, as Alan and his wife were expecting the family's first grandchild.

A.A. Albert's year at Yale was financed by the National Science Foundation, which had arranged for him to set up a "special year" on algebra. The "special year" convinced him of the value of such "special years," a concept he later brought to Chicago.[19]

Besides setting up the project, he actively participated in it. He lectured at a colloquium on finite fields, a topic that had attracted renewed attention due to the advent of that recent invention, the digital computer.[20] He highlighted the topic in his sixth textbook, *Fundamental Concepts of Higher Algebra*, published in 1956.

His "pure research" coalesced with his "applied" defense work when it came to certain problems in finite fields. He had produced a classified paper for the government on finite fields in 1955. The year he taught at Yale, his paper on "trinomial equations in finite fields" appeared in the *Annals of Mathematics*.[21]

Spending the year at Yale afforded him an opportunity to collaborate with Jacobson, who had managed to secure a teaching position at Yale in 1947 as discrimination against Jews at top universities subsided in the wake of World War II.[22]

By 1956, both men had acquired a keen interest in Jordan algebras. At the time, Adrian was fine-tuning his groundbreaking paper showing that "every simple exceptional Jordan algebra is either 'reduced' or is a division algebra."[23] While Adrian placed the finishing touches on his lengthy paper defining Albert division algebras,∥ the two men teamed up to write a paper on the other class of Albert algebras, that is, "reduced" exceptional Jordan algebras.[24] Their joint paper succeeded in filling major gaps in the research on "reduced Albert algebras."

Adrian remained in the East for the summer of 1957 to direct a six-week project in Brunswick, Maine sponsored by the Air Force Cambridge Research Center. He had invited some of the brightest lights in the field of modern algebra to summer with him in Maine. All of the mathematicians in the project received top-secret clearance in order to participate.

He was in his element that summer, surrounded by a number of his dearest colleagues and their families. His former doctoral students, Richard Schafer, Chair of the Mathematics Department at the University of Connecticut, and Louis Kokoris, a professor at IIT, were there with their wives. Lowell Paige and his wife Peggy arrived to join the ensemble, straight from the project at Yale. Irving ("Kap") Kaplansky rented a house for the summer with his wife and children. Kap delivered a crisp, charismatic lecture early in the session. I. N. ("Yitz") Herstein joined the project, along with Daniel Gorenstein, Erwin Kleinfeld, W. Mills, and George Seligman. Nathan ("Jake") Jacobson and his wife, "Florie," occasionally drove up from New Haven to visit.

During the week, the mathematicians split up into offices lent to them for the summer by Bowdoin College. The specific problems they worked on involved a type of cryptographic system the Air Force was investigating.[25] Although the Air Force designed the problems for specific applications, the researchers did not know how their work

---

∥ More on Adrian's discovery of Albert algebras and their division algebras in Chapter 16.

would be used. To the mathematicians, the problems appeared to consist of pure theoretical research.[26]

Many of the problems assigned by the Air Force related to group theory. Others were of a "combinatorial character."[27] "Combinatorial analysis" was becoming one of Adrian's specialties.#

There were weekend parties at a nearby salt water lake. At one such gathering, Adrian caught and cooked a 2 1/2 pound lobster which he served up at a lobster bake for his colleagues. Carrying his platter of steaming lobster, he beamed from ear to ear. He was the picture of a professor pleased as punch at orchestrating an idyllic summer retreat where like-minded mathematicians could "play" at what they loved best—pure mathematics.

By the end of the summer, Herstein and Kaplansky produced some significant results in group theory by showing that ABA-groups are necessarily solvable.[28] Gorenstein, who had focused on ring theory until then, discovered during the summer that his forte was group theory.[29] Gorenstein's shift to this specialty was fortuitous. Some years later, he would earn high praise for playing a crucial role in classifying finite simple groups—a major achievement of twentieth century mathematics.[30]

The year after Adrian's return from his glorious year in the east would be filled with both triumph and tragedy. Nineteen fifty-eight began with his election as Chairman of the Mathematics Section of the National Academy of Sciences, the country's most respected scientific body.

A month later, the University's Board of Trustees elevated him to the chairmanship of the Mathematics Department. After serving so many intermittent terms as Acting Chairman, it was time.

By then, he was somewhat conflicted about taking on this new set of duties, as he had so many other irons in the fire. Still, he accepted out of a sense of responsibility to the University and his colleagues at the Department.

On the other hand, assuming the position did not mean that he was chained to the University. He spent the spring quarter of 1958 as Visiting Professor at UCLA. During the ensuing summer, Adrian and

---

\# See discussion of combinatorial analysis and its relationship to cryptology in Chapter 22.

Frieda would travel to Europe for the International Congress of Mathematicians.

From the perspective of the graduate Department of Mathematics, the Hutchins plan was a dismal failure. Hutchins' liberal arts-oriented undergraduate mathematics program enabled students to read Euclid in the original Greek text. However, graduates of the College lacked a sufficient background in mathematics to gain entrance to the University's graduate Mathematics Department. They were prepared to become well-rounded Renaissance Men and Women of the 17th century, but ill-equipped to meet the challenges of the 20th century age of specialization.

The Mathematics Department recognized the need to bring the two programs into line. A few months after Adrian assumed the chairmanship, an agreement was reached dissolving the separate mathematics program in the College and incorporating it into the graduate Department of Mathematics.[31] Henceforth, the undergraduate mathematics curriculum would consist of a unified series of core courses designed to serve as building blocks for graduate studies.

As the size of the Alberts' apartment grew, so did the size of Frieda's parties. From a young bride totally lacking in culinary skills, she became a chairman's wife adept at cooking for a large crowd. On at least one occasion, she served an elaborate dinner to the Department's entire faculty as well as graduate students and their respective spouses—about 100 people in all. She paid close attention to tricks of the trade divulged by the country's most famous hostess, Perle Mesta.

A friend sent her a *New Yorker* magazine cartoon depicting a rather portly hostess bidding goodnight to her guests. One of the departing guests in the cartoon is shown remarking, "Heaven only knows how you construct these super parties, Frieda. It must take a lot out of you."

Her parties were not mere social events—they were a fusing of people into a mathematical community that fostered their scholarly work. She and Adrian often used the term "mathematical family." When new doctoral students or faculty members joined the Department, Frieda helped them find housing, furnish their homes, and generally, served as a one-person Welcome Wagon.

Tragedy waited until the fall of 1958. It happened on the fourth anniversary of Alan's "Golden Day," the day he was released, alive and uninjured, from military service. The news arrived without warning in a telephone call from California, where the Alberts' diabetic son, Roy, had stopped to visit Frieda's parents. Roy had suddenly contracted a gut-wrenching case of the flu. He underestimated the difficulty of maintaining a constant bloodsugar level in the face of a virulent flu bug. The macho kid from Chicago resisted his grandparents' wish to rush him to the hospital. A doctor came and left, but soon afterwards, his condition became critical. Within a space of 24 hours, death had claimed the Alberts' younger son. He was twenty-three.

Adrian and Frieda were devastated, and, although they never discussed it, the pain of their loss persisted the rest of their lives. Due to Roy's interest in studying anthropology, the Alberts established the annual Roy D. Albert Prize to benefit anthropology students at Chicago.

The blow was softened somewhat when, two months after Roy's death, first son Alan and his wife became the proud parents of their second child, a baby boy. The couple gave Adrian and Frieda naming rights. Befitting the circumstances, they chose the name "Roy."

At the same time, Adrian gave birth to the proof of another theorem. While pacing the hospital corridor awaiting his grandchild's birth, he had worked out the solution in his head.[32] It was a problem that seemed hopeless to him only a week beforehand. He wrote to a colleague at the Institute for Advanced Study, "I have determined the collineation groups of the finite planes associated with the algebras called twisted fields. The astonishing fact is that these groups are all solvable."[33]

Two of A.A. Albert's mathematical "sons" were Reuben Sandler and John Thompson. Adrian and his protégés often took long walks around East Hyde Park together, discussing solutions to mathematical problems.

He and his protégés shared a passion for mathematics. According to one account, Adrian had begun working on a joint project with Sandler, but needed to depart for a summer of work in California in the midst of their collaboration. Apparently, he thought about the project during his entire road trip from Chicago to California. He telephoned Sandler to discuss it at every stop along the way.[34]

John Thompson, a tall, gaunt young man, was one of the most brilliant mathematics students on campus in the late fifties. Although Adrian was not Thompson's doctoral supervisor, he took a fatherly interest in the young genius' development.

Thompson had developed a fascination for group theory. At the time, Adrian was intrigued by the problem of the minimum number of generators for algebraic structures, a problem that he had addressed in connection with his defense work.* In 1958, the synergy of their talents resulted in a joint paper concerning the generation of a collineation group by two of its elements.[35] Their collaboration grew out of a formerly classified research paper on the same topic[36] that Adrian produced for the Air Force Cambridge Research Center in 1956–57 for use in cryptography.†

When Thompson submitted a topic for his doctoral thesis, his thesis advisor, Saunders Mac Lane, discouraged him. Mac Lane told Thompson, "Don't work on finite groups. It's a dead-end subject." He assured the young man that the problem he proposed to solve was too difficult. "You'll never solve it," he insisted.[37] Thompson stuck to his guns and kept his topic.

In 1959, Thompson received a doctorate based on his thesis, where he proved a conjecture of Frobenius relating to finite groups. According to J. J. O'Connor and E. F. Robertson at the University of St. Andrews, Scotland, "The solution of Frobenius' conjecture was not done by simply pushing the existing techniques further than others had done; rather it was achieved by introducing many highly original ideas which were to lead to many developments in group theory....It is no coincidence that starting at the time of Thompson's thesis, group theory leapt into prominence as the mathematical topic which was attracting the most attention and which was undergoing the most rapid development. The reason was that suddenly progress began to be made on one of the main problems of finite group theory, namely the classification of finite simple groups."[38]

While Thompson's specialty differed from his own, Adrian saw Thompson in some sense as his mathematical heir. There was tangible

---

* A "generator" is a mathematical entity that yields another mathematical entity or its elements when subjected to one or more operations.

† In a letter to the author, Thompson confirmed that the problem was based on classified work done by Adrian prior to their joint paper. (Letter to N. E. Albert from J. G. Thompson, January 6, 2004.)

symbolism in the fact that, when it came time for the Alberts to move out of their Hyde Park Boulevard apartment, Thompson and his wife happily inherited the claw-footed dining room set that had hosted so many Mathematics Department dinners. Mrs. Thompson, who loved old things, had no doubt admired it during one of Frieda's famous dinner parties.

The most difficult of A.A. Albert's new duties as Chairman of the Mathematics Department was preparing the departmental budget. He agonized over the task. His expertise in higher mathematics did not automatically translate into turning a few loaves into enough fishes to feed and grow the entire Department.

Once in office as Chairman, he set to work on fulfilling a variety of goals. One of his priorities was seeking salary increases designed to keep some of the Department's top mathematicians from being lured away by other universities with fatter budgets. He was sorely disappointed when he lost his fight to secure adequate incentives for several of them. Professors Shiing-Shen Chern and Ed Spanier left Chicago for Berkeley and Halmos went to the University of Michigan.

Another priority was raising funds for some of his pet projects. Along with other members of a steering group he served on, Adrian secured a $30,000 grant from the Defense Department for a University research project to bring visiting mathematicians to campus.[39] He also sought out funding from major corporations and received a $30,000 research grant from the ESSO Education Foundation.[40] Among the grants that flowed into the University's coffers during his chairmanship was a National Science Foundation grant in topology.

The number of activities he undertook simultaneously in late 1958 and early 1959 boggles the mind. While teaching, conducting his own mathematical research, raising grant monies, and administering the Mathematics Department, he managed to complete a host of other tasks. He provided a litany of his activities to the administration that included: writing a research paper for the Golden Jubilee Celebration of the Calcutta Mathematical Society;[41] serving on the Executive Committee of the School Mathematics Study Group at Yale University, a project to rewrite texts for high school mathematics classes; acting as peace-maker in resolving a dispute between the University of Chicago Press and the Department of Statistics about publishing a series of monographs on statistical research; organizing and chairing a Symposium on Group Theory in New York City on behalf of the

American Mathematical Society; and participating in a series of mathematics conferences sponsored by St. Xavier College.[42] Afterwards, for a relaxing summer, he chaired SCAMP, a mathematics research project at UCLA dealing with top secret government research.

Funding for the Symposium in New York came from the Office of Naval Research and IDA's Communications Division. The impact of government-supported research became evident at the Symposium. A presentation by Gorenstein attributed significant progress in group theory to the seeds that were planted at the Air Force summer project on group theory at Bowdoin College in 1957. The summer project had spawned a considerable amount of later work in group theory by both Gorenstein and Herstein.[43]

There were a number of new faces at the Symposium. Three keynote speakers, John Thompson, Marshall Hall, and Michio Suzuki, called attention to new results in finite groups. Eleven speakers in all contributed papers later published by the American Mathematical Society.[44] The body of work presented at the Symposium further advanced the subject of group theory and set the stage for the breakthroughs achieved during the "group theory year" that Adrian was organizing for 1960.

Gorenstein became friends with Herstein during the Maine project and later hooked up with Walter Feit and John Thompson. In subsequent years, the four of them brainstormed together about problems in group theory.

One of the best perks of a life in academe was owning the summer months. For Adrian, summertimes were part work and part play. He had spent most of his summers in California beginning in 1950. A second set of his fishing gear took up residence at the Los Angeles home of Frieda's sister. Frieda relished the chance to spend a portion of each year in the Golden State with her parents and siblings.

His summer jobs amounted to a second career. As previously described, summertimes allowed him to conduct research in applied mathematics for government-sponsored projects like RAND and SCAMP. In the process, he developed friendships with mathematicians at UCLA and the University of Southern California. He had a close collegial relationship with Lowell Paige, with whom he coauthored a paper on Jordan algebras.[45]

Paige and others began an initiative to snare the Chicago professor for the permanent faculty at UCLA. Perhaps the administration at

Chicago saw it coming. They sweetened his pay package even before hearing about the negotiations. Upon learning about the possibility of his leaving, the Dean used his persuasive powers to convince him to stay. In the end, it became apparent that even the allure of Southern California could not draw him away from Chicago's Mathematics Department.

Kaplansky observed that, as Chairman, A.A. Albert left his stamp on the Department by "maintaining a lively flow of visitors and research instructors, for whom he skillfully got support in the form of research grants. The University cooperated by making an apartment building available to house the visitors. Affectionately called 'the compound,' the modest building has been the birthplace of many a fine theorem. Especially memorable was the academic year 1960–1961, when Walter Feit and John Thompson, visiting for the entire year, made their big breakthrough in finite group theory by proving that all groups of odd order are solvable."[46]

The "group theory year" at Chicago, like the "special year" on algebra at Yale, was underwritten by the Air Force's Office of Science Research and the National Science Foundation. In addition to Thompson and Feit, Adrian invited ten other exceptional mathematicians to join in the elite company. The most distinguished among the group was Richard Brauer, who had made significant progress in a concerted effort to classify all simple groups during the 1950s. Gorenstein arranged to spend his sabbatical at Chicago that year in order to participate.[47]

Chicago's "group theory year" had the trappings of a miniature Institute for Advanced Study. Researchers were free to concentrate entirely on pure research for the academic year, sans even a concern about housing. The results proved the value of the model. Commenting on the program several years later, Adrian described its significance: "The theory of groups is an important subject with applications not only to various aspects of mathematical thought, but also to chemistry and physics. The special program was enormously successful, and its achievements have been major ones in the theory."[48]

Through the use of grant monies, Chairman Albert had managed to attract some of the ablest mathematicians of the day to Chicago's campus. One of the mathematicians he attempted to hire was the brilliant John Nash, who, Adrian later learned, had succumbed to schizophrenia.[49] Adrian knew the unconventional genius from the time both men consulted for RAND in the early fifties.

According to Nash's biographer, Sylvia Nasar, Nash thanked Albert for his kind offer, but demurred that he would have to decline because he "was scheduled to become Emperor of Antarctica."[50] Nasar reports that the matter came to the attention of MIT's president, who, after seeing Nash's response, commented, "This is a very sick man."[51]

In a retrospective, former Chairman Marshall Stone praised Adrian Albert's "strong leadership" of the Department. He concluded that, under Adrian's leadership, the Department "flourished mightily and was able to maintain its acknowledged position at the top of American mathematics."[52]

Despite his position as departmental chairman, his authority to hire permanent faculty members was limited. In 1958, he had lost his bid to bring Nathan Jacobson on board.[53] Adrian was extremely disappointed, since he regarded Jacobson as one of the country's top algebraists. Although there were those who felt that religious affiliation was a factor, the reason given was that Jacobson's field of mathematics was not the one most needed in the Department at that time.[54] During the ensuing three years, a number of other mathematicians were hired or promoted, none being of Adrian's choosing.

In June 1961 he sought to secure the talented algebraist I. N. Herstein to fill the vacancy of Otto Schilling, who was leaving for an appointment at Purdue University. Adrian wrote a letter to "Dear Colleagues." In it, he described two letters that he was preparing to send the Dean. "In one I have recorded the vote for Herstein's appointment as unanimous and recommending his appointment....The other is my letter of resignation as Chairman of the Department...It is up to you to decide which letter I will send."[55] His strategy succeeded. The faculty approved Herstein's appointment.

Chapter 21

Urban Removal and Condomania

It was an annual rite of spring in the Albert household. Frieda invariably reserved a table for the Mother's Day Luncheon at the Quadrangle Club, the faculty meeting-place that had served as her husband's playground since 1931. We donned our best outfits, expecting to exchange pleasantries with many of the faculty families dining there.

The brick and limestone structure at 57th Street and University Avenue, with its domestic Gothic design, resembled a private English men's club. The main floor hosted a large billiards room with a stone fireplace. Nearby, the garden room overlooked the tennis courts. While catering to the University's mostly male faculty, the club offered a luxurious amenity for its female guests. Ladies entered a posh marble lavatory by way of a spacious lounge filled with overstuffed chairs and sofas. Every inch of the club, from its highly polished stone staircase to the marble floor on the second level, bespoke the elegance of a bygone era. The architecture of the faculty club blended well with the Gothic structures on campus.

The sixties would dramatically alter the architecture of the adjoining neighborhood. Back in 1954, the Chicago Land Clearance Commission[1] had succeeded in having portions of Hyde Park designated as "slum and blighted area redevelopment projects."* That designation, combined with 1954 amendments broadening the federal Housing Act to include slum prevention as well as slum clearance, laid the foundation for a building boom close to campus.

---

* See discussion of Hyde Park urban renewal in Chapter 18 above.

The Commission acquired nearly 43 acres of land to be cleared and sold to developers. The cost of clearance was borne by the government, effectively giving the developers a subsidy.[2]

By 1959, many of the buildings called for under the redevelopment plan were under construction. While several high rise apartment buildings were designed as rentals, a series of low rise townhouse subdivisions were planned. These attached homes would be deeded separately as single family residences. Due to the low cost of the land, the new homes would be relatively affordable, enabling many of Hyde Park's former apartment dwellers to become first time homeowners.

From a bird's eye view, the long rows of flat-roofed sand-colored townhouses resembled army barracks. The box-like affairs took the then popular Mies van der Rohe idea that "less is more" to an extreme. Their lack of ornamentation appealed to the psyche of the time. The popular mood reacted against Victorian styles in favor of open space and large expanses of windows.

Architecture aside, what attracted many Hyde Parkers to the new townhomes was their kitchen cabinetry, modern heating and cooling systems, and state of the art appliances. These were scarce amenities in a neighborhood that had seen little new residential construction for decades. Although the urban dwellings came with party walls, half a dozen alternative floor plans appeared generous compared to the tiny ranch homes built for veterans in suburban subdivisions during the fifties.

Hyde Park was on the brink of another metamorphosis in its housing market at that moment in time. A novel concept in real estate began to emerge.

Savvy Hyde Park landlords savored the expectation that the neighborhood was about to receive a massive infusion of real estate investment dollars. The apartments that had been generating only modest rental income stood to appreciate in value as middle class residents returned to the city.

The landlords faced a quandary—how to cash in on the regentrification of Hyde Park. The answer seemed to lie in selling off individual apartments rather than entire apartment buildings. Some of these landlords saw an opportunity to pass along the costs of deferred maintenance to a new set of owners. But few buyers could afford to purchase the apartments without bank financing, and still fewer lending institutions were willing to provide loans on individual apartments marketed in the usual way—as shares in a "cooperative."

Some creative lawyers came up with a solution to the landlords' dilemma. They devised a strategy to make apartments as easy to finance as single family homes. These enterprising attorneys concocted a legal fiction whereby an apartment assumed a dual identity as a percentage of one or more buildings and an entity unto itself. Borrowing the concept of joint dominion from Roman law, they called the new entities "condominiums."

Landlords soon discovered that, with the aid of some lawyers and fancy paperwork, they could reap huge profits almost overnight by converting their apartment buildings into condominiums. In 1959, the spacious rental apartments lining Hyde Park Boulevard and other streets in East Hyde Park began their conversion. It was the dawning of "condomania" in Chicago.

The Alberts' luxury Hyde Park Boulevard apartment building was one of the first slated for conversion. The building's owner presented them with two stark alternatives—purchase an apartment, or be evicted.

The Alberts considered their options. It was easier to remain with all of their worldly possessions in the sprawling second-story apartment. But, given their advance into late middle age, they were enticed by the opportunity to acquire reserved parking at their doorstep, a ground level entry, and a low maintenance home with modern amenities. They opted to buy one of the affordable new townhomes under construction in the "golden triangle" close to campus.

They selected an end unit on the alley with a diminutive backyard. Adrian ordered wrought iron gates opening onto the alley. He had the yard paved with a concrete parking pad for his Opel, leaving just enough soil around the periphery for some ornamental bushes and a few flowers. No grass for him to mow—it was perfect! There were no garages—the family car would occupy a reserved space in front of the house.

The picture window in the living room overlooked a wrought iron backyard fence affording no privacy. Frieda loved the beautiful firethorn bushes she had seen dotting the landscape in Princeton. She persuaded her landscaper to plant them along the fence, despite his admonition that they would not withstand Chicago's harsh winters. Should the firethorns survive, she would be able to view their spectacular red berries from the picture window.

Their townhouse faced a brand new privately owned street created by the subdivision's developers. The developers named it "Park Place," although the concrete enclave was devoid of greenery save for some back yard plantings provided by the first batch of homeowners and an unadorned swatch of grass where the subdivision abutted 55th Street's busy thoroughfare.

The subdivision gained immediate popularity among Hyde Park's intelligentsia. Homeowners would include three mathematics professors along with other university faculty, physicians, lawyers, accountants, and a network news anchorman.

The old apartment was good for one last party. My fiancé and I had been engaged since March 1959, but had yet to set a wedding date. In early August, it occurred to us that we might start the school year as a married couple. Frieda was away visiting relatives in California when we came up with the idea.

True to form, she rushed home from California and sprang into action. She contacted an engraver, had the apartment spruced up, hired a caterer, made arrangements with a florist, and took me shopping for a wedding dress.

Theatre-style seating filled the parlor of our Hyde Park Boulevard apartment in September 1959. My fiancé and I walked down the makeshift aisle past a crowd of well-wishers. Afterwards, the claw-footed dining room table, laden with hors d'oeuvres, performed its final duty for the Alberts.

Chapter 22

E. H. Moore at the Entry

From my earliest memories, an autographed reprint of Eliakim Hastings Moore's portrait hung in the foyer of our 57th Street apartment. Garbed in a maroon taffeta academic gown topped by a hood trimmed in gold velvet, the first Chairman of Chicago's Mathematics Department watched over me during all of my comings and goings as a young child.*

Now, it was two residences and nearly a decade later. The Alberts were moving to Park Place. Before settling into their new home, Adrian and Frieda left to spend the summer of 1960 in Los Angeles, where he would direct Project SCAMP, the Southern California Applied Mathematics Project financed by the Defense Department.

The timing was perfect for my husband and me. We had spent the past year in a furnished, off-campus apartment while attending graduate school. Like brother Alan and his wife, we moved in to house-sit while awaiting our first child.

In July, the following letter from the University of Chicago's Chancellor Lawrence Kimpton reached Adrian in California: "It gives me great pleasure to inform you that I am today appointing you Eliakim Hastings Moore Distinguished Service Professor. It is a very great honor and no one could deserve it more."[1] The chaired

---

* The original portrait of Moore was painted during construction of Eckhart Hall with the understanding that it would hang in the University's first mathematics building. A collection was taken up among Moore's friends, colleagues, and former students to fund the portrait. Approximately 300 individuals contributed to the fund. Color prints were made to send to each of the donors.

professorship, one of only thirteen at Chicago, was particularly meaningful to Adrian because its first occupant had been his Chicago mentor, Leonard Dickson.

While Dickson was one of his chief mentors, Adrian saw Moore as his role model. From the moment they met, Moore made an indelible impression on the young student. Years later, Adrian recalled, "During my first year as a graduate student I came under the influence of E. H. Moore who had much to do with the abstraction of my later work."[2]

He admired Moore for more than his mathematics. He revered the man's humanity and his ability to advance mathematics by shaping institutions. Like Moore, he recognized that undertaking administrative duties, while consuming valuable research time, afforded an opportunity to improve both his university and his profession.

The Mathematics Department at Chicago owes a great deal to Moore. After serving as acting head of the Department for four years, he became chairman in 1896. In all, Moore stood at the helm of the Department for 35 years and built it into a powerhouse of pure mathematical research.

Apart from Moore's crucial role as the founding Chairman of the Department, his leadership was impressive. He was instrumental in transforming the American Mathematical Society from its origins as the New York Mathematical Society into a national organization. He then managed to shift some of the concentration of mathematical activity from the east coast to the Midwest by spearheading a Chicago section of the American Mathematical Society, notwithstanding vigorous opposition and criticism on the part of some east coast mathematicians.[3]

Moore was elected AMS President in 1901 and delivered the Society's Colloquium lectures in 1906.[4] He was the central moving spirit behind the founding of the Society's prestigious publication, *Transactions of the American Mathematical Society*.[5]

His influence extended beyond mathematics. Moore was recognized as a leader throughout the scientific world. He became President of the American Association for the Advancement of Science in 1921.

The larger-than-life leader inspired his former students to speak of him in superlatives. When he died in 1932, Harvard mathematician George Birkhoff wrote to Moore's widow, "What I particularly admired in his so lovable personality was his complete devotion to the highest ideals which was of a degree of purity...rarely to be found. On account of this union of mathematical greatness with utter fairness, generosity, and breadth of outlook, his students not only admired but held him in affectionate admiration.... I am so glad to possess the reproduction of that lovely portrait which I often look at since it seems to bring me into his very presence."[6] Dickson wrote in his obituary of Moore that "he was a leader who was universally loved, and this was because he was...a prince of a man."[7]

A.A. Albert became the third person to hold the Moore chair, the second recipient being a microbiologist named William Taliaferro who left the University to work at Argonne National Laboratory. The local newspaper heralded the appointment with the headline "Mathematics gets its chair back."[8]

Congratulations poured in from colleagues and acquaintances, near and far. Upon news of the honor, Harry Everett, the Alberts' former Hyde Park Boulevard neighbor, penned this note: "Hearty congrats on EHM Dist. Sv. Prof!!! Hooray. Ya beat Mac to it!!! Don't let Stone propose a swap!!!"[9] For his part, Saunders Mac Lane graciously sent his "hearty congratulations."[10]

The Alberts' former 57th Street neighbor, Daniel Boorstin, wrote to express his pleasure. "In the last couple of years I have traveled a bit to remote parts of the world lecturing about American History. In the course of these travels I have been pleased to discover that wherever anyone has heard of the University of Chicago and has any interest in Mathematics they know you and your work. It has pleased me to think that your fame has reached South Asia!"[11]

Alan Simpson, Dean of the University of Chicago's college, sent over a whimsical note. "I have just returned from a vacation to learn that Dickson's pupil is now sitting in his chair. Warmest congratulations."[12]

The Defense Department acknowledged the appointment with the tongue in cheek quip, "We want you to understand, however, that this does not entitle you to a raise in pay or an increase in travel allowances. The only obvious result will be a slightly more humble demeanor on the part of the Panel Executive Secretary in your presence."[13]

Along with the honor came an increase in salary. Adrian would receive $20,000 per annum—the top salary paid to any mathematician at Chicago. The increase was a welcome contribution to the Alberts' first venture into home ownership.

In September, the Alberts returned from California to their new home. They arrived just in time to greet their brand new grandchild. Adrian's fondness for cameras came in handy. He snapped photos of his new grandbaby with me, with my husband, and even with the baby nurse. He could not get enough photos of his grandchild. Frieda's wallet grew fatter. She referred to herself as a SOGWPIP (Silly Old Grandma With Photos In Purse).

Having given away most of the old furniture, she had a field day replacing it. He redoubled his efforts to supplement their income with consulting work.

She purchased two new armchairs for the living room and placed them facing the picture window that overlooked the backyard. There would be no more formal dining room for large Mathematics Department dinners. Instead, the new floor plan offered a dining area near the kitchen. She replaced the heavy, claw-footed dining furniture with some Lane pieces of Scandinavian design. Faculty dinners would be relocated to a large "recreation room" in the windowless basement.

The squat exterior of the new townhouse was deceptive. Inside, it was surprisingly spacious, with three full stories, including the basement.

Although empty nesters, the Alberts required space for visiting children and grandchildren. They chose a five-bedroom model. Adrian and Frieda now had his-and-hers studies. She took a room opposite the kitchen and arranged to have one wall lined with bookshelves, while he chose one of the upstairs bedrooms for his study and had two floor-to-ceiling bookcases made for his ever-growing collection of whodunits. Like his Princeton mentor Wedderburn, he was endlessly fascinated by the brain-teasing tales.

Although pure mathematics is considered a young person's game, Adrian showed no signs of tapering off. During his fifties, his mathematical research ranged over a variety of topics, from associative algebras and matrix theory, to Jordan and other nonassociative algebras, to combinatorial analysis, twisted fields, and projective planes. He declared the subject of projective planes to be "a rich and rewarding area of mathematics, whose study uses a wide variety of techniques—geometric, algebraic, combinatorial, etc."[14]

He published several papers on projective planes.[15] Adrian viewed "twisted fields," which he had invented in 1952, to be "an important chapter in the theory of finite projective planes since every twisted field is a coordinate ring for a non-Desarguesian plane."[16] The concept of "isotopy" that he had introduced in connection with his study of nonassociative algebras turned out to be "exactly what was needed in studying collineations of projective planes."[17]

In a departure from his usual subject matter, he refuted a construction of non-Desarguesian planes by the mathematician T. G. Ostrom. Ostrom later referred to Adrian's article in two papers of his own.[18]

He believed some abstract mathematical topics to be within the grasp of high school students. In a lecture to a group of secondary school teachers, he suggested that they consider adding a course on projective planes to the high school curriculum in order to give young students a sense of what pure mathematical research is all about.[19]

His work demonstrated the symbiotic relationship between algebra and geometry. In his college text, *Solid Analytic Geometry*, he provided an algebraic vector approach to the study of geometry.[20] The book ties up the study of space analytic geometry with the theory of vectors and matrices, and leads into the topic of projective geometry.

A significant portion of his work dealt with "combinatorial analysis." The topic incorporates a variety of disciplines, including coding theory and cryptology. Little has been written about this aspect of his work because much of it was classified.

He delivered a paper on the topic at a Symposium on Combinatorial Analysis sponsored by the Army's Office of Ordnance Research in April 1958.[21] He later addressed the subject of combinatorial analysis in a text on finite projective planes that he coauthored with his former doctoral student, Reuben Sandler.[22]

What portion of A.A. Albert's papers in "pure" mathematical research has been exposed to the light of day remains an open question. Seven of his books and approximately 143 of his mathematical papers were eventually published; three, posthumously.[23]

Due to the sensitive nature of his defense-related work, many of his papers remain unpublished. Some of his previously classified manuscripts on defense research were later "sanitized" to read like papers in "pure" mathematical research. Approximately 30 of these manuscripts have been declassified, and, with the Government's permission, he had a few of them published. For example, in 1954, the National Security Agency agreed to allow him to publish a "Report on Finite Fields" that he had furnished the agency under a contract, provided that he mention only the Department of Defense, and not the

National Security Agency.[24] In 1967, Adrian published an article on polynomial systems that he originally produced as a Working Paper for SCAMP.[25]

His colleague and close friend at the Chicago's Mathematics Department, Izaak Wirszup, reported that Adrian never spoke about his top secret work. Wirszup was quick to point out that this was particularly hard on him because one of his greatest joys in life was discussing his work with colleagues.[26]

Most of his defense work was housed where the work was performed, in secure facilities.[27] A mathematics professor who consulted for the Institute for Defense Analyses during the 1960s recalled that IDA housed approximately 100 of Adrian's research papers on its premises. The professor observed, "He blazed a trail in applied algebra that led many others to follow in his footsteps."[28] The majority of these papers remain classified to the present day.

Undoubtedly, his research in the realm of theoretical science informed his applied mathematics at projects like RAND, SCAMP, and IDA. However, scientists who worked alongside Adrian on these projects believed that his success in applied mathematics was not due to his training alone. They attributed his ability to solve practical problems to his innate capacity for abstract analytical thinking and his insight into identities and calculations.[29]

A more controversial question is whether, and to what extent, his years of defense-related research enhanced his research in "pure" mathematics. Some who knew his work believed that results in applied mathematics did not lend themselves to thinking about pure mathematical research. On the other hand, many of the problems he solved at government-sponsored projects were presented in the form of pure research questions. In such cases, his classified work consisted of pure research.

In response to a survey of scientific personnel, Adrian listed the four major areas of his expertise in the following order of competence: (1) "rings, fields, and algebras;" (2) "linear algebra and matrix theory;" (3) "combinatorial analysis;" and (4) "finite geometries."[30] Jacobson observed that, while A.A. Albert considered himself a pure algebraist, his best work—the solution of the problem of multiplication algebras of Riemann matrices—had its origins in geometry. Moreover, he "could exploit analytic and number theoretic results when he needed them...."[31]

Chapter 23

An Entirely New Dimension

After only a year in their new home, the Alberts were off again. This time, Adrian was slated to spend the 1961–62 academic year in Princeton as Director of the Institute for Defense Analyses' new Communications Research Division.

Perhaps it was only a coincidence. The last time he spent an academic year away from Chicago, the University promoted him to head the Mathematics Department.

Shortly after taking his leave, the man who began his journey in one of Chicago's poorest immigrant neighborhoods received word of his appointment as Dean of the Physical Sciences Division at Chicago's premier university. He would henceforth be responsible for the Departments of Astronomy, Chemistry, Geophysical Sciences, Physics, and Statistics as well as Mathematics. In addition, the job involved overseeing the Institutes for Nuclear Studies, the Study of Metals, and Computer Research.

The appointment was a seminal event in two respects. He crashed through old barriers by becoming the first University of Chicago Physical Sciences Dean of Jewish ancestry. He had broken a similar barrier only three years earlier as the Mathematics Department's first Jewish Chairman. Secondly, it served as a tacit acknowledgment by faculty members of the Division who voted for him that a researcher of pure mathematics deserved to head the entire group of physical sciences. Dickson's belief that mathematics is the Queen of all Sciences was playing out in reality.

Early in his career, Adrian had perceived that interest in pure mathematics was waning in America. Mathematics was widely viewed as a rather esoteric pursuit, and jobs in the field were scarce. Some thirty years later, mathematics appeared to be flourishing and had greater application than ever to the other physical sciences.

He was the man of the moment for a variety of reasons. His own credentials as a leader in pure mathematics gave him the credibility necessary to interact with his peers in the Division. His defense-related consulting had afforded him the chance to interact with scientists of other disciplines and learn their needs. He was well known around the University for his contacts within government and private industry that brought an infusion of research grants into the school. At the Department of Mathematics, he was considered an excellent administrator. And, perhaps not a factor to be minimized, he treated others with respect.

His appointment reminded faculty members of the former Dean whose apartment had perched directly above the Albert residence in the 1950s. Of all prior Physical Sciences Deans at Chicago, Walter Bartky's academic background most closely resembled that of the new Dean. Bartky earned a doctorate in mathematics at the University of Chicago in 1926. He taught in the Department of Astronomy until 1942, when he switched over to teach applied mathematics, mathematical astronomy, and statistics in the Mathematics Department. Bartky took over the deanship in 1945.

Bartky played a key role in advancing the status of statistical science at the University. In 1949, he carved a Committee on Statistics out of the Mathematics Department. The Committee eventually evolved into a full-fledged Department.[1] Bartky died in 1958 at the age of 57, three years after stepping down as Dean.

Well-wishers greeted the news of Adrian Albert's appointment with enthusiasm. Warner Wick wrote from the viewpoint of the University's College, "we will miss you and your knack for multiplying a few loaves and fishes in our plans for mathematics."

Former Mathematics Department Chair Marshall Stone sent his regards from Santiago, Chile: "The Division is fortunate in securing you as its Dean. Congratulations all around! Of course this will create some problems in the Department, but it should eliminate lack of understanding in the Dean's office!"[2]

One of the most touching congratulatory notes came from a former student. "You perhaps do not remember me for I was just one of a myriad of your fledgling math students at the University...In my judgment there could have been no more deserving man...I wanted to drop you a line...and to let you know that...every student has a favorite professor—you are mine."[3]

Another former student wrote, "You bring a scholarly distinction to the office which not only adds luster to it but guarantees that you will give proper consideration to the scholarly problems of your colleagues. Now if only you really like the administrative work and can do it without interfering with your mathematical research, all will be perfect."[4]

Harry ("Sheidy") Everett sent another of his colorful notes. "Now this here Deanery business is outta this world. In fact, us faithful co-ugh-workers just simply dogedly do *not* believe it! It *is* just a gag, isn't it?...Wotta long-suffering line-up: Salisbury, Gale, Bartky, Johnson, Zachariasen, Albert!! You do realize, don't you, that they all die or fold up in the deanery?"[5]

Adrian received encouragement from other colleagues at Chicago's Mathematics Department. Professor Herman Meyer wrote, "The job is as important as it is tough; but if you're as good a Dean as you were a Chairman (and of course you will be), you'll be the best damn Dean these hallowed halls have ever seen."[6]

The wife of his colleague, Alberto Calderon, had a unique way of celebrating the appointment. She telephoned Frieda, who avoided gardening due to her allergies. "A Dean's wife should have roses in her garden," she announced. Whereupon she arrived in the tiny yard behind the Alberts' townhouse, spade and rosebushes in hand.

Perhaps Leonard Blumenthal best summed up Adrian's feelings: "Congratulations or commiserations (whichever you find appropriate) on your Deanship. If you can do *that* and mathematics too, you will be setting some kind of a record."[7]

His fans in Buildings and Grounds celebrated in their own way. A mustard yellow wastebasket appeared in his office emblazoned with large block lettering that read "Office of the Dean."

His appointment to the Deanship began on January 1, 1962, but he remained in residence at IDA through the remainder of the 1961–62 academic year. He would retain the post as Dean until his mandatory retirement at age 65 in 1971.

Blumenthal's concern that Adrian's new responsibilities might stand in the way of his pure mathematics research may have been mitigated by his record after becoming Dean. He continued to produce research, despite his administrative duties, albeit at a reduced pace.

During his term as Dean, he published at least a dozen mathematical papers dealing with such diverse topics as "Jordan algebras," "Riemann matrices," "finite planes," "biabelian fields," "cyclic algebras," and "polynomial systems." In one of these papers, he advanced his prior research into "Albert algebras" by constructing a new class of "noncyclic exceptional Jordan algebras" that he referred to as "bicyclic algebras."[8]

Returning to his 1930s research into "central division algebras of prime degree," he poked and prodded them to determine whether any were noncyclic, and developed some new results.[9] The decades-old issue of whether these algebras must be cyclic is a tantalizing question that continues to baffle mathematicians.[10]

He also edited an anthology of articles titled *Studies in Modern Algebra*[11] and published revised versions of his texts, *College Algebra* and *Solid Analytic Geometry*. Following up on his analytic geometry book, he expanded his explorations into the interplay between algebra and geometry by suggesting a collaboration with Sandler, who specialized in projective geometry.[12] In their 1968 treatise, they show that "[o]ne of the beautiful features of the study of finite projective planes is the way in which it connects certain concepts of algebra with those of geometry."[13] The subject of "projective planes," they note, is, like number theory, within the grasp of the mathematical novice.

As soon as he returned from Princeton, Adrian set to work with gusto on beefing up the Physical Sciences Division. In many respects, he was a man ahead of his time. He recognized the importance of computers early on.

Early in his administration as Dean, he wrote to the National Science Foundation that he had "managed to shepherd through our Division a Committee...on Information Science. This will probably mean that I will have in the Division a Department of Information Science."[14] The letter was a prelude to a proposal for NSF seed money to support the Division's computer effort.

He proceeded methodically to pave the way for the new Department by advising the University's administration, "the Institute for Computer Research for the Committee on Information Sciences...is expected to grow rapidly and to become a seventh department in the Division."[15] He made the following pitch to the University for more money to grow the fledgling "Committee:" "Assistantships and fellowships will be needed for what is certain to be a rapidly growing scientific discipline which will provide strong support for a number of other disciplines in the University. Several new appointments will be needed in this Committee before, what we regard as its critical size, will be attained."

Adrian was not new to computers. He was involved in top-secret work on finite fields when the earliest digital computers were being developed. When asked to speak on "The Modern Computer" at the National Youth Conference on the Atom in 1962, he began by professing a "lively ignorance" of the subject. Whereupon he proceeded to describe the various types of computers invented since the beginning of time, the names of their inventors, how modern computers functioned in both narrative and mathematical terms, what their component parts were, which computer languages were extant, the speed and power of the new machines, and potential applications for computers in the future.

As Dean, A.A. Albert had once more to reinvent himself. Although he had administered the Mathematics Department, heading the entire Physical Sciences Division was a task of another magnitude entirely. His new duties as head of one of the University's four divisions encompassed administering an annual budget of $10 million and overseeing a faculty of 120.[16]

The *Chicago Tribune* described the new Dean as "a mild-mannered man who chooses his words as carefully as he handles the symbols with which he created the science of Jordan algebras...."[17] Adrian gave the newspaper his bird's eye view of the University. "We are distinguished by the emphasis which is placed on research. The rank of a university in the modern world depends on its distinction in scientific accomplishments."

He saw the graduate Division of Physical Sciences as one of the freest communities in the world. In his view, "Freedom is having the time to do research...Work never leaves you. It is something you don't give up—ever.... The work is of the man's own choosing. He is not told what to do....Of course,...[e]ven in mathematics there are 'fashions.' This doesn't mean that the researcher is controlled by them. Many go their own way, ignoring the fashionable. That's part of the strength of a great university."[18]

With regard to the teaching function of a great university, he told the *Tribune*, "It is not generally recognized by the public that the best teacher is the person who knows...This may not be sufficient to make him a good teacher, but it is necessary. It is the first requirement.... It is necessary that you have people who themselves contribute to knowledge. That you obtain fine teachers is incidental."

The new Dean soon became aware of an altogether different dimension in the administrative obstacles to be hurdled. In a 1963 address to a group of college "plant administrators," he summed up his latest challenge: "the main problem of the Physical Sciences Division is one of space."[19]

After two years as Dean, he submitted a ten-year plan for revamping the Division. He began by assessing the problem: "The Division's eminence is no longer as clear as it was at one time." The problems, he suggested, were due in large part to the emergence of other great new centers of science in the country and the fact that "[t]he bad reputation of our neighborhood before urban renewal still persists, and makes more difficult the recruitment of top faculty and top students." The goal—"We expect to strengthen the Division during the next ten years to a point where it is obvious to our competitors and to graduate students that we are at least as good in all of our areas as any institution in the country."

New buildings were among his priorities. He observed, "The only really modern laboratory...is the Space Laboratory, a part of the Fermi Institute." With the help of a committee he appointed in 1962 to study the Division's building requirements, he set forth a proposal for new buildings to house chemistry, high energy physics, and expansion of other departments. The plan also proposed a new Science Center where maximum interaction between the sciences could take place, and which included connecting a biochemistry and biophysics complex to a joint physical and biological science library.

In 1963, he announced plans for a $4.5 million chemical research laboratory and restoration of the aging Jones and Kent chemical laboratories at a cost of $1 million.[20] He explained that the new laboratory would be "a unit in the long-range science center now being planned by the university."[21]

It fell to Adrian, as head of the University's Computer Policy Committee, to administer a new $500,000 grant from the National Science Foundation for support of the school's Computation Center. The Center was founded in 1962 with the aid of IBM and served all of the disciplines throughout the University.[22]

In his 1965 report to the University's administration, he reported that most of the funds had already been raised for the high-energy physics building and for the first part of the Science Center that would house geophysical sciences. He capped off his report with the prediction: "With vigilance and aggressive recruitment by the faculty, and excellence as our goal, the Division cannot help but have a golden future."

While he attended to building up the Physical Sciences Division, he remained the man to see about fine-tuning government grant proposals for projects within the Mathematics Department. Marshall Stone wrote to him in September 1965, "When you can find time after clearing up the urgent matters which will inevitably greet you on your return, I would like to have a short discussion with you about a research proposal I want to put in with AFOSR....What I need to discuss with you is the coordination of this proposal with the other elements of my program. I would, quite naturally, like to see the University and myself get the full benefit of the contract as finally written, and this may require a bit of care."[23]

If Adrian's skill at funneling federal grants into University research projects was a motivating factor behind his selection as Dean, he did not disappoint. At a convocation address delivered shortly before stepping down as Dean, he reported that, for the 1971 fiscal year, the Division had received 123 grants and contracts from the federal government totaling nearly $10.5 million.[24]

To the extent that the pace of his pure mathematics research slowed during that period, it was as much attributable to his responsibilities outside the University as to his in-house duties. Despite the additional duties as Dean, the scope of his extracurricular duties actually accelerated.

Throughout the 1950s, he had been seriously overloaded with outside responsibilities such as advisory posts at the National Research Council and National Science Foundation. During his years as Dean, his obligations beyond the Quadrangles seemed to increase exponentially. The added workload might have crushed a healthy man, let alone one beset by diabetes and recurring colitis.

He needed his occasional fishing trips to regenerate his energies. For Adrian, fishing was as much medicinal as recreational. With his extended duties, he rarely had time for them.

He never stayed in one spot for long; he seemed to be everywhere at once. When not on the Chicago campus, he was either attending a ceremonial function, lecturing at a university, traveling to Europe for mathematical meetings, attending board meetings at government agencies or academic institutions, or leading technical assistance missions on university campuses.

During the fall of 1963, he attended the Centennial Celebration of the National Academy of Sciences commemorating the one hundredth year of its founding. It was a grand affair, with an opening address by President John F. Kennedy and a formal procession of Academy members and foreign associates dressed in academic costumes. Three days of Scientific Sessions were held at the State Department.

One of the speakers at the Scientific Sessions was Eugene Wigner, who had sparked Adrian's interest in an "algebra of quantum mechanics" in Princeton 30 years earlier. Wigner garnered the Nobel Prize for physics in 1963. Although Wigner served on the faculty at Princeton, he had conducted much of the research for which he received the prize while working on the Manhattan Project at the University of Chicago.[25]

Wigner benefited in one important respect from the fact that his specialty was mathematical physics, and not pure mathematics. It is widely believed among mathematicians that Alfred Nobel, founder of the Prize, harbored a deep enmity for Gosta Mittag-Leffler, considered the father of Swedish mathematics. As the story goes, Nobel deliberately deleted mathematics from the list of subjects eligible for his prize to guarantee that Mittag-Leffler would never receive it.[26]

As Dean, he concerned himself with the prospects of doctoral students and faculty members throughout the Physical Sciences Division. When the Chairman of one Department washed his hands of a young black student who had failed to curry favor with that Chairman, Adrian wrote to the Department head, urging that he go out of his way to help the minority student secure a good position. Upon receiving the Dean's letter, the Department head readily agreed to help the young man.

In the mid-1960s, a professor of physics and chemistry named Robert Mulliken came to Dean Albert for help.[27] The aging scientist feared that his appointment contract would not be renewed. His daughter required expensive medical attention and he needed the income. Would Adrian please see what he could do to keep him on the payroll and allow him to finish his research?

Professor Mulliken was up against the University of Chicago's mandatory retirement policy. Faculty members were automatically retired at the age of 65 unless the University administration opted to continue their service. As a rule, after age 65, reappointments were for one year only; however, upon recommendation by the President and the Board of Trustees, the University could grant a three-year deferred "retirement appointment."

Before the ban on age discrimination in employment became the law of the land, the term "retirement appointment" was not considered an oxymoron. It was on account of the University's retirement policy that some of Adrian's former professors had moved on to less prestigious institutions that welcomed these scholars to their faculties with open arms.

Adrian went to bat for Mulliken and succeeded in securing a contract renewal for him. He thought no more about it until November 1966. Mulliken had taken a leave for the quarter to teach at Florida State University and so could not be reached at his home in Chicago.

It turned out to be Dean Albert who awoke to the nerve jangling ringing of the telephone at about five o'clock in the morning. Once he heard the speaker on the other end of the line, his irritation at the intrusion on his slumber vanished. It would be one of the most joyous telephone conversations of his life. The Swedish Royal Academy of Science was calling from Stockholm. The Academy had voted to award the 70-year old Mulliken a Nobel Prize for his work in molecular orbital theory.

Molecular orbital theory, which describes the path traveled by electrons in complex molecules, helps scientists predict the behavior of molecules by looking at individual atoms. The *Chicago Sun-Times* quipped, "University of Chicago Prof. Robert S. Mulliken, who once climbed mountains as a hobby, reached the summit of his profession Thursday when he was awarded the Nobel Prize for chemistry."[28]

The process of gaining a promotion in academe is not unlike rising in any other type of organization. One supplicant who wrote seeking Adrian's support described the process as follows: "Now that I've been 'academic' for a year, the departmental deacon agrees that the time has come to nominate me for elevation from associate to full sainthood. Part of the ritual seems to involve accumulating testimonials from saints and prelates in other parishes to impress the local bishop. Favorable evidence should be addressed to…. Evidence to the contrary should, of course, be addressed to the Devil's Advocate…."[29] The supplicant was amply rewarded with a letter from

Dean Albert ending in the magic phrase "I strongly recommend his promotion."[30]

Having consulted in private industry and served with some of the leaders of industry on various boards, Adrian knew where the jobs were in mathematics. He advised the young Horst Feistel, who felt his interest in exploring cryptography was being stifled at the National Security Agency, to seek employment at IBM. Feistel later flourished in IBM's intellectual playground, developing ways to protect privacy in cyberspace.[31]

In 1964, A.A. Albert followed in the footsteps of E. H. Moore and L.E. Dickson by becoming President of the American Mathematical Society.[32] The AMS had come to play a primary role in furthering mathematical research since Moore shepherded the fledgling organization into national prominence during the preceding century.

The following year, Notre Dame University awarded Adrian an honorary Doctor of Laws degree, noting that he "has boldly extended the limits of his special field—notably as one of the principal developers of the theory of Linear Associative Algebras and as a pioneer in the development of Linear Non-associative Algebras."

Chapter 24

Turmoil on Campus

The 1960s witnessed a revolution in American academia far more dramatic than the quiet revolution that took place after World War II. In the wake of the war, Jewish professors slowly gained acceptance on the nation's campuses. During the 1960s, America's colleges and universities once again confronted questions of race and ethnicity. The problem predominated in the nation's Southland, where skin pigmentation often determined a student's access to education.

President John F. Kennedy's main priority in 1963 was fighting the cold war. As more African nations became independent, they sent their ambassadors to the United States and established consulates in and around Washington, D.C. The President, who needed the support of these countries at the United Nations, was embarrassed by the housing discrimination encountered by the new ambassadors.

At the same time, the President's brother, U.S. Attorney General Robert Kennedy, toured the southern states and was deeply moved by the suffering of black Americans.[1] He heard about southern blacks being chain-whipped to prevent them from exercising their right to vote, refused admission to public libraries, and turned away from "whites only" public grade schools and state-run universities.

In 1954, the United States Supreme Court in the case of *Brown v. Board of Education* had considered the constitutionality of segregation in public schools. The High Court ruled that the Constitution's guarantee of equal protection required all publicly supported schools to admit blacks inasmuch as separate educational facilities were inherently unequal.

In his first run for the Alabama Governor's mansion, George Wallace declined support from the Ku Klux Klan. As a result, Wallace lost the election to a man who ran on a racist platform and accepted the Klan's support.[2] Wallace learned from the experience that it was necessary to cater to segregationist views in order to win the State House. In his 1962 campaign, Wallace pledged that, if elected governor, he would flout any federal court orders to integrate the state's schools.

In June 1963, the U. S. government filed suit against Governor Wallace to enjoin him from preventing admission of two black students to the University of Alabama.[3] When the two students arrived to enroll at the University, the Governor himself stood in the doorway, blocking their entrance while an angry crowd looked on.

Attorney General Kennedy ordered 500 Alabama National Guard troops to the University campus. He also sent his Deputy Attorney General, Nicholas Katzenbach, whose towering presence conveyed a powerful message. The governor backed down, and the students entered safely.

President Kennedy lacked the legal tools to end the pernicious state-supported practices of discrimination and segregation. By using the National Guard to stand up to southern governors, the Executive Branch appeared to be going out on a legal limb to enforce a court-made rule. Calling in the National Guard every time a black student attempted to enter a state-supported school was neither a practical nor a popular solution to inequality under the law.

The federal courts possessed no means to enforce their own mandates, since they controlled no army or militia. It was unclear how southern states could be made to comply with federal court rulings.

The televising of the University of Alabama confrontation between Wallace and Katzenbach was one of the catalysts that changed public opinion to the point where President Kennedy felt empowered to seek a legislative solution to the problem.

The day after the confrontation, the Ku Klux Klan continued its reign of terror, determined to intimidate and even murder "uppity" Negroes who dared to speak out about their oppression. On June 12, 1963 the respected civil rights leader Medgar Evers was shot down in the driveway of his Mississippi home while his children cringed in fear inside.

The nation at large shared the black community's outrage. With the wind of public opinion at his back, the President knew that the country was ready for legislative reform. At Medgar Evers' funeral, President

Kennedy handed Mr. Evers' widow a copy of the proposed Civil Rights Act that would give the United States government a new set of tools to fight discrimination and injustice against minorities.

Within a week of the assassination, Kennedy delivered his bill to Congress in support of sweeping civil rights legislation. Kennedy's proposal included a guarantee that no employer could discriminate against an individual on the basis of race or national origin.

Southern congressmen were especially antagonistic to the proposal. But when Kennedy was cut down along a Dallas parade route only a few months after the bill had been introduced in Congress, there was a groundswell of support for the bill.

The nation went into deep mourning after Kennedy's assassination on November 22, 1963. Upon being sworn in as President, Lyndon Johnson sought to sooth the pain of the American public with the words "Let us continue." Johnson would keep his word. He not only continued the civil rights struggle, he would achieve a civil rights victory that might have been impossible for his northern predecessor.

The President from Texas placed his substantial political capital behind pushing the Civil Rights Act through Congress, alienating many southern congressmen along the way. Using every bit of the "clout" he had accrued as a former southern senator, Johnson succeeded in achieving the bill's passage. Congress tacked the new 1964 Civil Rights Act onto the 1866 federal statute that had created basic rights for newly freed slaves at the end of the Civil War.[4]

Johnson saw the new law as a means of eliminating barriers to employment. It was part and parcel of Johnson's "War on Poverty" initiative—a step toward establishing "The Great Society." A "Great Society," Johnson wrote, is one that "rests on abundance and liberty for all. It demands an end to poverty and racial injustice."[5]

The civil rights proposal was not the only campaign that Johnson inherited from his former running mate. President Kennedy believed it to be in America's national interest to help defend the non-Communist government in South Vietnam against armed incursions from the north. The U.S. government had armed and supported South Vietnam's despotic President, Ngo Dinh Diem, expecting him to resist Communist aggression. Instead, Diem wielded his power to repress Buddhist monks and other political opponents.

In the eyes of some observers, President Kennedy's involvement with the November 2, 1963 overthrow and assassination of Ngo Dinh

Diem may have been a turning point in what became no longer a Vietnamese but an American war.[6]

During Kennedy's presidency, the United States positioned 16,000 soldiers in South Vietnam, the majority of whom served in administrative and intelligence positions. In addition, U.S. Air Force and Army pilots were attacking suspected Viet Cong targets. Close to eighty American soldiers died during fighting in Vietnam before President Johnson took office.[7] Most Americans were unaware of the extent of America's involvement in Vietnam during the Kennedy administration.

The Gulf of Tonkin incident, which ostensibly triggered America's bombing of North Vietnam, occurred on President Johnson's watch in August 1964. Believing that supporting the government of South Vietnam was necessary to prevent the spread of Communism, the hawkish new president sought, and received, a resolution from Congress giving him authority to "prevent further aggression." Johnson stepped up war. Over a period of several years, he gradually sent hundreds of thousands of American troops to Vietnam as U.S. casualties mounted.

The period following the United States' sending of combat troops into North Vietnam in March 1965 was a painful one for America. Students around the country began massive protests as the war dragged on.

Protesting students at the University of Chicago targeted everyone they believed to have any connection to the U.S. war machine. A.A. Albert would not be exempt. The University acceded to the demands of the Students for a Democratic Society to investigate whether the school should disaffiliate from the Institute for Defense Analyses on the ground that it was part of the Vietnam War effort.

In November 1967, the *Chicago Maroon* displayed Adrian's photograph prominently on the front page next to the headline, "University-IDA Tie Comes Under Fire." The paper noted his role as a trustee at IDA and quoted him as stating that "he felt it was his patriotic duty to work for the defense of the United States.... [and that] even if the University were to withdraw from IDA, he and other professors would continue working for them..."[8]

The same issue printed this statement by a leader of Students for a Democratic Society: "The University should withdraw from IDA...It has no right to act as a legitimizing agency for the most blatant attempt of the military to impose on academia. It is an aid to American politico-military hegemony over the world, which we oppose."[9] The contrast in student attitudes between the Vietnam and World War II eras could not have been starker.

A large group of University of Chicago students, fortified by shopping bags full of groceries, staged a lengthy sit-in at the Administration Building. Later, they surrounded the home of the University's President, Edward Levi, marching in protest. There were some tense moments, not knowing whether the crowd might become violent, or what could happen next. Levi waited the students out. Eventually, they tired of marching and left on their own accord.

At one point, the students and their shopping bags even invaded that citadel of gentility, the Quadrangle Club. They brushed past the Club's staff, who sat cowering inside the reception area, feeling defenseless against the students' brazen intrusion. But the students were no match for Frieda Albert. She happened by, and, upset at seeing the students with their untidy groceries and too casual poses lounging around the elegant lobby, screamed at them, "This has been my living room for the past 35 years." Whereupon, she delivered a soccer kick that landed one of the grocery bags clear across the room. You didn't want to tangle with Frieda Albert when she had her dander up.

The terrified students quickly evacuated the building. The receptionist at the front desk looked at Frieda and smiled. "Thank you so much. We didn't know what to do."

In 1968, the University faculty caved in to the students' demands by voting to withdraw from IDA. Adrian told the *Washington Post*: "My feeling is that IDA has developed an academic atmosphere so that it is strong enough to attract the people it wants," and he would continue as a trustee because he served in an individual capacity. He added that there had never been the slightest hint that he should drop the trusteeship, even from faculty members who have been the more vocal critics of foreign policy, particularly regarding Vietnam.[10]

Student protests continued to disrupt the work of the University for the remainder of the decade and on into the early 1970s. In a February 1969 letter to a colleague, he apologized for his tardy response. "My delay in responding is due to the many meetings I have been attending since the sit-in began."[11]

Chicago's problems were minor compared to those of some other universities. Some of the most violent disturbances took place in Ohio.

During the winter of 1969–70, there were angry clashes at Ohio State University between students and law enforcement authorities that caused classes to be cancelled and the campus, closed.[12]

The following spring, students at Ohio's Kent State University staged protests and engaged in acts of vandalism in the center of town. The Kent city mayor declared a state of emergency and telephoned Ohio's governor, asking for help.[13] The governor dispatched the Ohio National Guard to the scene.

The next day, students torched a barracks housing the Army Reserve Officer Training Corps on campus and prevented firemen from dousing the fire by damaging their equipment. During the ensuing two days, National Guardsmen tried repeatedly to disperse the crowd.

On May 4, 1970, two thousand people gathered on the Kent State University commons, flouting an order barring any further rallies. The guard ordered the crowd to disperse, but the crowd replied with chants, curses, and rocks. The confrontation continued to escalate after the firing of tear gas canisters failed to disperse the crowd.

Using fixed bayonets, the guard forced demonstrators to retreat to a nearby practice field. Once on the practice field, the guard saw that the crowd had not dispersed and that the field was fenced on three sides, leaving the guardsmen surrounded.[14] The rock-throwing and verbal abuse continued, unabated.

As the guardsmen withdrew, demonstrators followed close behind. As they approached the crest of a hill, guardsmen turned and fired toward the parking lot. When the firing ceased, four students lay dying and nine others were wounded. A University ambulance drove through the campus, announcing that the campus was officially closed and ordering all students to evacuate the campus at once.

Chapter 25

Education Guru

Adrian possessed a rather dry sense of humor. In a 1963 Convocation address to graduating students, he recalled his own graduation. "I remember now how uncomfortable was my seat, but I cannot remember either who gave the address or what was said. Strangely enough, I find that thought very comforting."[1]

Dean Albert was dedicated to eliminating mediocrity in the nation's institutions of higher learning. He preached, "The production of a single scientist of first magnitude will have a greater impact on our civilization than the production of fifty mediocre Ph.D.s."[2]

During November 1964, he spoke at the University of South Florida in Tampa at a conference on "Science Today and Tomorrow" convened to dedicate the school's new Physics-Astronomy-Mathematics Building. Adrian delivered a thoughtful speech entitled "The Curriculum in Mathematics" on changes in college mathematics education over the past forty years.

He described how, shortly after World War II, he and then Chairman of Chicago's Mathematics Department, M. H. Stone, needed to engineer a total revamping of the mathematics curriculum in order to develop a unified curriculum leading to the master's degree. To produce students having a sufficiently broad knowledge of mathematics, he and Stone increased the number and types of courses required. Adrian recalled the poor quality of many of the mathematics textbooks then in use and how he had to totally rewrite the syllabi while seeking out improved texts as they began to appear.

Dean Albert wrapped up the lecture with a report on how far mathematics education had come and how far it had yet to go. He pointed out serious gaps in the required course lists of many institutions that left students without a proper foundation to continue their studies.

Noting the recent development of curricula leading to higher degrees in Applied Mathematics, he indicated that, while he approved of those degrees, he believed that such programs too often eliminated certain core "pure mathematics" courses needed to understand the field.

He concluded, "the present curricula in most institutions show a great increase in the breadth and depth of coverage of modern mathematics over what existed 40 years ago. In particular, modern algebra has been introduced into the program of undergraduates in most places. What is lacking is a study of the whole program, an appreciation of the unity of mathematics, and the structure of a program which will place particular units at the proper place.... [T]he order in which courses are given is important as well as the courses themselves..."[3]

A.A. Albert's perspective as Dean and mathematics curriculum guru heralded a series of new consultancies. He was sought after for technical assistance and evaluation services in developing and upgrading mathematics and science programs around the country.

At home, Illinois Governor Otto Kerner appointed him to membership in a state Science Advisory Council.[4] At the national level, he served as consultant to the U. S. Office of Education, the predecessor agency to the Department of Education.

He became a man with a mission—the Johnny Appleseed of mathematics. His goal—to spread his ideas about the best way to teach mathematics as far and wide as possible.

Depending upon the situation, he traveled alone or led teams of experts called in by universities around the country to help with curriculum development or to provide advice in getting new facilities off the ground. Much depended on the recommendations in these site visit reports. A favorable report could result in seed money for new institutions from the Science Development Program of the National Science Foundation. *Sputnik* had caught America off guard in 1957 and NSF was pouring resources into the race for scientific supremacy.

In January 1965, Adrian conducted a review for Rice Institute regarding their proposed transformation into Rice University, and penned the team's report with suggestions for implementing the plan and expanding the faculty. While critiquing Rice's proposal as the product of "haste and much which is in the realm of pious hope," he concluded that Rice would be a top candidate for NSF support.[5]

Later that year, he made a site visit to Purdue University, again, under the auspices of NSF's Science Development Evaluation Group. He followed up the Purdue site visit with a five-nation South American speaking tour where he met privately with university and government leaders who sought his advice on their mathematics curricula.

The following year, he met with officials at the University of Iowa for two and a half days to assist in developing a new Institute for Applied Algebra. His former doctoral student Robert Oehmke, a mathematics professor at the University of Iowa, was one of the driving forces behind the proposed Institute. A portion of Adrian's time on-site was devoted to discussing potential funding by charitable foundations for various aspects of the Institute proposal.

Soon after completing his technical assistance mission in Iowa, he was asked to assist with the organization of a proposed Institute for Fluid Dynamics and Applied Mathematics at the University of Maryland. He headed a team of three university professors who visited the site in September 1967 and filed a report with recommendations for the new Institute.

The next university to seek his input as curriculum consultant was the University of South Florida, where he had lectured in 1964 and again in 1966. The main purpose of his May 1969 visit was to assist the University's Mathematics Department in initiating its first Ph.D. degree program.[6]

A technical assistance request came from Ohio the following year. In 1970, the administration at Ohio State University and Arnold Ross, Chairman of its Department of Mathematics, asked Adrian to visit the school with a team of consultants to conduct a needs assessment. Ross, a former Dickson protégé at Chicago, had in mind positioning Ohio State's Mathematics Department to move up in the ratings of the American Council on Education.

Ross apologized for his delay in issuing the invitation, explaining, "As you know, we had a very difficult time on our campus last year. Our difficulties boiled over into national prominence late in the winter and that did not seem to be a propitious moment to ask our friends and colleagues to take a calm and searching look into the role of mathematics on campus."[7]

Adrian made a four-day visit to Ohio State in late March 1971 with a team of three other mathematicians and drafted a report afterwards addressing the needs of the Mathematics Department. The report pinpointed one of the essential differences between a state school and a private college.

As part of a state system of colleges, Ohio State was obliged to provide services that a private school like the University of Chicago did not. Ohio State had instituted a "Head Start" program to counteract the inadequate preparation of students coming from Ohio's less privileged communities. The team found that, while "Head Start" was essential, it overtaxed the faculty's workload. The school's broad mandate also consumed resources that might otherwise be used to recruit top notch faculty members. Due to the steep hurdles that a state school faced in meeting budgetary requirements to comply with all of the recommendations in the report, Ohio State would have to ease into the proposed upgrades gradually.[8]

Besides his technical assistance visits and foreign travel, Adrian was stretched to the limit with trips involving his role as advisor to various schools, governmental bodies, and private industry. For instance, in 1968, after returning from lecturing in Puerto Rico, he accepted a three-year reappointment to the Advisory Council of Princeton University's Mathematics Department.

The same year, he was elected to membership in the American Academy of Arts and Sciences, the nation's second oldest learned society.[9] The organization was founded in 1780 by John Adams and other members of the Massachusetts Bay Colony. Adrian did not take any of the honors bestowed upon him lightly. The son of the impecunious Russian immigrant was still striving to succeed.

Dean Albert was generally on the list of invitees for the dedication of any new science facility in the country. He was among those helping to turn first soil at a groundbreaking ceremony for a new Science Center at the Belfer Graduate School of Yeshiva University in October 1965. The following month, he spoke at a New York City Conference on "Some Recent Advances in the Basic Sciences." He attended the dedication of the Heritage Building at M.I.T. in December 1967 and, in 1970, participated in dedication ceremonies for Princeton University's new Fine Hall.

Yeshiva University chose to honor him in April 1968 with an honorary Doctor of Science degree for his "continuous and enduring contribution to the advancement" of his chosen field. The Jewish university presented him with an oversized diploma worded entirely in Latin. He hung it on his wall alongside the diploma, inscribed in English, presented to him in 1965 by the University of Notre Dame.

In the midst of his serious endeavors, the Chicago professor found time for a bit of levity. He attended Chicago's 1968 Mathematics Club skit "Bwana John in the Outback" featuring students impersonating faculty members such as Adrian ("Father Abraham"), Saunders Mac Lane ("MacFlame"), Felix Browder ("The Unbounded Operator"), and John Thompson ("Bwana Thompson").

Midway in the skit, the narrator announced, "Here's father Abraham...." Adrian's impersonator made his entrance wearing erasers tied to both elbows and began writing with both hands and, occasionally, with his feet, before speaking.

Father Abraham delivered his lines: "Alas, I'm lost trying to find a left-handed piece of chalk. And my son Isaacs is no better off. He's abandoned the mini-group-theory seminar—become an idol-worshipper, searching for a shrine to the great god Brauer. And ah, the poor children of Israel, also known as the Fellowship of the Ring—they've been wandering through the rings for forty years, trying to find him between here and Rome." At the end of the skit, Browder's impersonator called out, "While Abraham is up on the mountain perhaps he can get me a grant."

A.A. Albert's career came full circle in 1969 with his election to the Institute for Advanced Study's Board of Trustees. His role as Trustee came over 35 years after his arrival at the Institute in 1933 as one of the first batch of young temporary members.

He became the lone academician serving on the 15-member board, which was dominated by leaders in industry, government, and law.[10] Adrian revered the renowned think-tank that had played a major role in America's pursuit of scientific exploration since the 1930s.

However, he soon discovered that all was not perfection in that Princeton paradise. Faculty members approached him complaining that their "consultative committee" was not in fact consulted by the Institute's administration before making faculty appointments. The administration's apparent disregard for their wishes fanned the flames of discontent among faculty members. He attempted to mediate the dispute but was assured that the faculty's perception was incorrect.[11]

That was not the only problem. The scientists on the Institute's faculty balked at the Board's domination by non-academics. The Chairman of the Board, an attorney, attempted to appease the faculty's testiest member, Adrian's former colleague André Weil, in a conciliatory letter. The attorney explained to Weil, a French citizen, that citizen participation in nonprofit organizations is one of the strengths that distinguishes a democracy like the United States from other countries. The irate professor shot back, "I have sometimes wondered how…[a lawyer] would have liked to have every major aspect of policy in his law firm submitted for final decision to a board of mathematicians."[12]

Adrian saw that the stalemate was hopeless. There was no way to change the composition of the Board, since it was embodied in the Institute's bylaws. Moreover, the Institute depended in large part upon private donations for its support. In a 1971 letter written shortly after Frieda's mother died of cancer, he wrote to the embattled Chairman of the Institute's Mathematics Department. He apologized for the fact that he would be unable to attend the forthcoming meeting between faculty and trustees to help mediate the situation and concluded, "I'm sorry that the Faculty unrest continues. I wish I had a magic wand to wave but I don't."[13]

The faculty of the Institute clung to his role as the only academic then on the Board and his reputation as a tactful conciliator. Weil later wrote of Adrian, "Deane [Montgomery] and I had great hopes that his position as a trustee might do much good to the Institute, or at least prevent much harm. In this, as in so much else, he cannot be replaced."[14]

Chapter 26

Building Bridges across the Globe

Adrian and Frieda Albert considered the community of mathematicians, both within the United States and throughout the world, as a family. They visited in the homes of mathematicians during their travels across the globe and reciprocated when foreign mathematicians visited Chicago.

In October 1962, at the height of the Cuban missile crisis, Russian nuclear missiles poised on Cuba's shores aimed directly at the United States. Americans were highly alarmed about the nuclear threat lurking only 90 miles off the coast of Florida. President John F. Kennedy, fearing that Russia would try to deliver more nuclear supplies to Cuba, took to the airwaves via television. He informed the American public that their Commander-in-Chief was setting up a naval blockade of Cuba to prevent any ships from reaching that island nation.

America's U.N. Ambassador, Adlai Stevenson, confronted the Soviet Ambassador at the United Nations with an aerial photo of Russian missiles perched on the island. When the Soviet leader refused to respond, Stevenson reacted with an atypical show of bravado. He announced, "I will wait for your answer until Hell freezes over."

At that critical juncture, Olga Olenik, a mathematics professor from the University of Moscow, was visiting Chicago. Both the Chicago mathematicians and their Russian visitor knew full well that their respective countries were on the brink of an all-out nuclear war. Had war broken out during her visit to America, the Russian guest was in peril of becoming toast along with her American counterparts.

Chicago's mathematicians greeted her warmly and hosted the Russian professor at faculty parties with no hesitation. The topics of politics and world affairs were pointedly avoided at these gatherings.

Toward the end of the professor's visit, Frieda brought her guest to the Alberts' townhouse and left her alone in the living room while Frieda went upstairs for a few minutes. By the time she returned, the professor had fallen into a peaceful slumber. "You can tell she's a lovely person," Frieda said afterwards. "She had such a sweet expression on her face while she was asleep."

For over a century, mathematicians have attempted to reach across national borders and political divides in the interest of science. They have not always been successful.

The success of the "Chicago Mathematics Congress" held in association with the the 1893 World's Fair set a precedent for international cooperation in mathematics. The neutral nation of Switzerland played host to the first official International Congress of Mathematicians (ICM) in 1897. Beginning in 1900, the international mathematical community sought to hold a Congress once every four years, each time in a different country. It was at the world's second ICM that David Hilbert famously delivered a talk on open "Mathematical Problems," offering up an agenda for the coming century.

World War I caused a schism between mathematicians whose countries fought on opposing sides during the war. Allied nations withdrew from existing international scientific organizations. In 1920, mathematicians from Allied Powers and neutral nations banded together to form the International Mathematical Union (IMU).

The IMU assumed responsibility for organizing the quadrennial congresses that had predated its existence. However, controversy erupted almost immediately over the fact that the IMU had adopted an explicit policy barring IMU membership by mathematicians from ex-enemy nations.[1] Membership in the world body was considered a prerequisite for attendance at the Congresses.

At the 1924 Congress in Toronto, American delegates urged the Union to reconsider its restrictions on membership. In the debate over which nations could attend Congresses, the IMU knocked heads with its parent organization, the International Research Council. The Council overruled the IMU and invited Germany, Austria, Hungary, and Bulgaria to attend the 1928 ICM held in Bologna, Italy. German mathematicians were widely recognized as the world's leaders, and excluding them from the Congress would have been a loss for the world mathematical community.

By 1928, Hilbert, the grand old man of German mathematics, had grown frail following a period of ill health. When he marched into the Bologna Congress at the head of a German delegation, a deafening silence descended upon the hall. Then, impelled by a common emotion, the sound of applause filled the chamber as the entire body rose to give him a standing ovation. The IMU's will was clearly not the will of the majority. The exclusionary attitude marginalized the IMU, which disbanded when its statutes expired in 1931.

The Congress of 1936 proceeded without the aegis of the IMU. Afterwards, organizers began planning for the first International Congress of Mathematicians to be hosted in the United States. Adrian received an invitation to deliver a one-hour "stated address" at a session on algebra at the 1940 Congress. However, when Germany invaded Poland in September 1939, all plans for international scientific cooperation were stilled. Upon hearing the news, an American benefactor who had pledged funds for the event opined, "It is too bad to have such a useless thing as a war interfere with such a useful thing as a mathematical conference."[2] No further Congresses would be held until 1950.

During his year as Visiting Professor in Brazil and Argentina, Adrian warmed to the role of goodwill ambassador. When the IMU was revived after World War II,* the American mathematician who had demonstrated a flair for diplomacy in the 1930s became very active in the world organization.

Having cancelled the 1940 Congress due to the outbreak of World War II, members of the world's mathematical community picked up where they left off after a ten-year hiatus. Since many of the donations collected for the 1940 meetings had been held in trust, it made sense to convene the 1950 Congress in Cambridge, Massachusetts where the 1940 Congress was to have met.

The Organizing Committee decided to invite about 20 mathematicians to deliver hour-long "stated addresses" to share the most significant recent developments in mathematics with attendees. Approximately 40 would be invited to give half-hour lectures, and there would be a good many 15-minute talks. Adrian Albert was delighted to assist in planning the number and types of conferences for the event.

---

* The first General Assembly of the new IMU was held in 1952.

The 1950 Congress began on a troubling note with the announcement: "Mathematicians from behind the Iron Curtain were uniformly prevented from attending the Congress by their own governments...."[3]

On the opening day of the Congress, Adrian presided at the Conference in Algebra where Zariski, Weil, MacLane, Garrett Birkhoff, Reinhold Baer, and C. Chevalley gave short presentations. During the fourth day, Adrian delivered a one-hour stated address on "Power-associative algebras" at a session on "structure theory of rings and algebras" where Brauer, Jacobson, and Jean Dieudonné also spoke.[4]

After 1950, IMU organizers selected European sites for the next few Congresses, beginning with the 1954 Congress to be held in Amsterdam. Frieda Albert accompanied her husband on most of his foreign travel. She was as outgoing as he was quiet and reserved. Together, theirs was a winning combination with respect to making and cementing friendships with mathematical couples in other countries.

The trip would be the Alberts' first visit to Europe and they were excited about the prospect. The Chicago professor would play a dual role as delegate to the General Assembly of the IMU and participant in the Congress.

The new IMU's General Assembly met in the seat of the United Nations' International Court of Justice, the Hague. Frieda recorded arriving at the end of August. "We were enchanted with the city. It is very old and charming yet modern and clean. Half the population is on bicycles and the other half on boats on the canals. Flowers are everywhere....Buses drove the delegates and wives to Delft for a reception at the Technical Institute."

During the meeting, tensions erupted between those who considered themselves "pure" mathematicians and applied mathematicians. Adrian was obliged to act as peacemaker in a rather nasty dispute between Marshall Stone, who represented the interests of pure mathematics, and F. J. Weyl, J. Tukey, and S. S. Wilks, representing the interests of applied mathematics.[5]

From the Hague, they proceeded on to Amsterdam, where Adrian chaired a delegation to the Congress on behalf of the National Academy of Sciences. The high-powered delegation also included Stone, Mac Lane, Adrian's former classmate Edward "Jimmy" Mc Shane, John von Neumann and Deane Montgomery of the Institute for Advanced Study, John Rosser of Cornell University, Carl Hille of Yale, and John Kline of the University of Pennsylvania.

Von Neumann delivered a Hilbert-like stated address early in the Congress titled "On Unsolved Problems in Mathematics." Later, Brauer reported on his research into the structure of groups of finite order in an hour-long address. Adrian chaired a lengthy session on algebra, which included a half-hour presentation by Jacobson on Jordan algebras.[6]

The trip was pure pleasure for Frieda and a great deal of work for Adrian. Frieda wrote, "We stayed at the Amstel Hotel—very luxurious. The Blumenthals, Wilders and von Neumann stayed there...Thurs. night there was a reception at the National Museum for all the visitors and families. The delegates to the International Math. Union & families were ushered thru a separate entrance and we were received by the Pres. of the Dutch Congress and he introduced us to the Burgomeister and his wife and the Minister of Education and his wife. These were flanked by guards in native costume of old in knee breeches with Napoleonic hats and red staffs."

The Alberts returned to Europe four summers later for the 1958 Congress. They traveled to Edinburgh and St. Andrews, Scotland, where Adrian again served as a National Academy of Sciences delegate to the IMU's International Congress and General Assembly.

He presided at a session on algebra and gave a short talk with L.J. Paige on "Jordan algebras with 3 generators"[7] while Frieda toured museums and ancient ruins. In the evening, Scottish hosts threw a party at a castle for the delegates. She noted in her diary, "The formal invitation was 'The Lord Provost, Magistrates and Council of the City of Edinburgh request the honor...at a garden party within the grounds of Lauriston Castle....' The Fireman's band played in full dress Scottish regalia. There were Scottish dancers at the big reception and dance."

Their third trip to Europe was for the 1962 Congress. Dean Albert headed a delegation on behalf of the Institute for Defense Analyses to the Congress held in Stockholm, Sweden. Once again, the elder statesman was designated to chair a session on algebra. To his delight, John Thompson delivered a half-hour address relating to his results on finite groups, while several former Albert doctoral students, Block, Osborn, and Zelinsky, each gave 15-minute talks on algebra.[8] American mathematics took center stage at the Congress when Princeton University's John Milnor gave a stated address on manifolds and received one of the coveted Fields Medals.

On the way home, the Alberts stopped off in Israel to visit with their son Alan who had a job with the Israeli aircraft industry that year. During their visit, they took the time to visit the tomb of the iconic figure, Theodore Herzl, whose identity had loomed so large in the communities where they grew up.

Adrian had no plans to attend the quadrennial International Congress set for Moscow in 1966. This was likely due to security concerns. Nevertheless, he contributed to the Congress by chairing the Committee to Select Speakers in Algebra for the meeting.

His next venture abroad was a return trip to South America on a Cultural Exchange program. Being an optimist about his time constraints, he prepared for the visit by investigating the rules and regulations for fishing in Southern Argentina's "fisherman's paradise."

In September 1965, he embarked on a 15,000-mile, five nation tour as a "Short Term Distinguished Lecturer." He lectured first in Brazil, then in Argentina, and capped off the tour with lectures in Chile, Peru, and Mexico.

Part of his mission on the tour was to advise academic and government leaders about ways to improve their graduate educational system. At a meeting with Argentine President Arturo Illia, he discussed the training of specialists in pure and applied mathematics and the use of electronic computers.[9] Afterwards, in his report to the administration at Chicago, Adrian described the mathematics curricula in the five countries he visited as being "in great ferment."[10]

During his stay in Argentina, he paid a visit to the laboratory of atomic beams at a research Institute under the auspices of the Comision Nacional de Energia Atomica. The laboratory, which benefited from an NSF grant, was engaged in research on measuring the charge equilibrium of helium beams.

He struck up a friendship with the director of the Institute, who had plans to visit Chicago's laboratory of molecular beams at the Enrico Fermi Institute. The director, an overcommitted scientist like A.A. Albert, later commented, "I remember your remark about the importance of being able to do many things at the same time and switch over rapidly. I realized it works."[11]

The mathematical community in the countries on Adrian's tour route went all out to show their appreciation by heaping honors upon him. He was named honorary member of the Mexican Mathematical Society, the Argentine Mathematical Society, and the Argentine

Academy of Sciences. The National University of LaPlata, Argentina named him an honorary member of the faculty, while the Brazilian Academy of Sciences voted to make him a Foreign Member.

While visiting Argentina, he saw his old friend Alberto Gonzalez-Dominguez, whom he had met during the post-war South American trip. Noting the instability of the political situation, Adrian asked the man whether he might be interested in coming to the United States. In 1966, Arturo Illia's government was overthrown in a military coup. The new military regime seized the country's nationally chartered universities.[12]

Adrian undertook a letter-writing campaign to his contacts in mathematics departments across the United States on behalf of the Argentine professor. Gonzalez-Dominguez was surprised at the flurry of activity once his North American colleague launched into action on his behalf. He wrote, "You did not forget; and shortly after your coming back to Chicago, a rain of invitations began to fall upon me."[13] The South American refugee arrived in the United States soon afterwards; he was among the quiet exodus of Argentine scholars following the coup.

The Chicago professor never received a warmer welcome than when he traveled south of the border. In March 1968, he ventured to the island of Puerto Rico, where he presented two lectures at the Universidad de Puerto Rico. To show their appreciation for his journey, Adrian's hosts presented him with a memento of the visit consisting of photos and clippings inside a folder emblazoned with a gold seal over green and white ribbons.

Programs for quadrennial congresses of the International Mathematical Union were the product of planning that began years ahead of time. In 1967, Dean Albert was named chairman of the "Consultative Committee" to plan the algebra program for the 1970 Congress.[14]

He flew to Lisbon and Paris for planning meetings in the fall of 1969, and later, to Stockholm, Sweden in the spring of 1970. Afterwards, a fellow member of the Consultative Committee recalled how, at the 1969 meeting in Paris, Adrian played the role of peacemaker when he, "with superior cleverness, presided over a difficult and sometimes quarrelsome assembly."[15]

Apart from his own travels as good will ambassador, Adrian helped to screen other potential candidates for foreign travel. One of his assignments was to interview a married couple applying for a cultural exchange program between the United States and the USSR. The National Academy of Sciences provided him with a list of questions to explore regarding the candidates' familiarity with the Russian language and customs and the couple's ability to endure physical hardships while in the Soviet Union. Whether or not he bore some of those issues in mind when preparing for his own visit to Russia some years later is an open question.

Rapprochement between the Soviet Union and western democracies came in small increments. In the spring of 1970, America's National Academy of Sciences extended an olive branch by electing the outstanding Russian mathematician, Israel M. Gelfand, as a Foreign Associate. By 1970, Gelfand had already distinguished himself in the areas of functional analysis, commutative ring theory, group theory, differential equations, integral geometry, and finite dimensional Lie algebras, and had made important contributions in the field of cell biology as well. Adrian, who regarded Gelfand highly, was delighted to learn of the Academy's decision.

Perhaps out fear of defection, the Soviet government seldom allowed mathematicians from the Soviet bloc to travel abroad for mathematical conferences. Adrian wrote to a fellow IMU member about the appointment, stating that he hoped, but did not believe, that the Soviet government might change its policy and allow Russian mathematicians to travel freely.[16]

The culmination of A.A. Albert's career came during his trip to Europe for the 1970 IMU meeting. At the end of August, he went to Menton, France as Delegate to the IMU's Sixth General Assembly, where he presided at the final meeting of a committee he chaired to select speakers for the 1970 International Congress.[17]

At the assembly in Menton, delegates elected Adrian to the position of Second Vice President of the IMU. His term of office was slated to extend from 1971 through 1974. As the only American office-holder in the IMU, it was a singular honor. Having previously served as President of the American Mathematical Society, he was now in line for presidency of the world organization.

From Menton, the Alberts proceeded to Nice for the ten-day Congress where Adrian's fingerprints were all over the program. Six of the speakers presenting papers at the Congress hailed from the University of Chicago faculty, including John Thompson, Jim Douglas, Jr., and George Glauberman. The local newspaper touted the event with the headline, "Pour Dix Jours, Nice Est Devenue La Capitale Mondiale Des Mathematiques." ("For Ten Days, Nice Has Become the World Capitol of Mathematics.")[18] The paper displayed the photograph of a beaming Adrian Albert shaking hands with Professor Laurentiev, First Vice President of the IMU. It was one of the American professor's proudest moments.

The trip to France was not exactly hardship duty. Monsieur et Madame Albert were wined and dined at the Palais d'Elysée and the Palais de la Mediterranée.

Adrian's unbounded pleasure at his election to the IMU post became doubly sweet when his protégé John Thompson received a Fields Medal at the Congress in Nice for his work on finite simple groups. At the IMU's invitation, Richard Brauer rose to give his imprimatur to the occasion. Brauer, whose important work on classifying finite simple groups helped to pave the way for Thompson's achievements, spoke with authority when he noted,

> The central outstanding problem in the theory of finite groups today is that of determining the simple finite groups. One may say that this problem goes back to Galois....up to the early 1960s, really nothing of interest was known about general simple groups of finite order.
>
> ...The first paper I have to mention is a joint paper by Walter Feit and John Thompson.... Here, the authors proved a famous conjecture, to the effect that all non-cyclic finite groups have even order....Fifty years ago...nobody had any idea how to get even started....It was only after the Feit-Thompson paper that one could be sure that the whole question was a reasonable one.
>
> Thompson's work which has now been honored by the Fields Medal is a sequel to this first paper. In it, he determines the minimal simple finite groups...
>
> These results are the first substantial results achieved concerning simple groups...
>
> [Thompson's] methods have already been used successfully by other mathematicians.... In this way, Thompson has had a tremendous influence. Since he first

appeared at the International Congress in Stockholm 8 years ago, finite group theory simply is not the same any more.[19]

Professor J. D. Fields, a Canadian mathematician, donated funds for the prize named in his honor. The Fields prize, regarded as "The Prize" in the field of mathematics, is restricted to mathematicians under the age of 40 because Fields wanted the prize to encourage future achievement. The first Fields Medals were awarded at the 1936 Congress. Afterwards, time stood still for the Fields prize while the mathematical world awaited the next Congress. The second set of Fields Medals was handed out at the 1950 Congress.[20] The long interruption in awarding the prize resulted in disqualifying many deserving mathematicians.

Adrian was extremely busy during the balance of the year. He managed to squeeze in a lecture at the Université D'Ottawa toward the end of 1970. During the fall, a number of universities played host to the President of the Academy of the Socialist Republic of Romania, who was visiting the United States under the auspices of America's National Academy of Sciences. The visit was part of an ongoing reciprocal exchange of scientists between the United States and Romania in accord with mutual agreements for scientific and cultural cooperation.[21] Professor Miron Nicolescu and his wife would spend three weeks in the United States meeting with American mathematicians.

Dean Albert hosted Professor and Mrs. Nicolescu during their visit to the University of Chicago campus in October. Afterwards, he received word from the National Academy that the Nicolescus "were particularly overwhelmed by the number of outstanding people you made it possible for them to meet."[22]

The following spring, Adrian was scheduled to meet briefly with Professor Nicolescu in Zürich, Switzerland. The National Security Agency was concerned about Adrian's safety anytime he was abroad, especially when he was about to meet with anyone from the Soviet bloc. On the eve of his departure for Zürich, he received a visit from an NSA security man, who asked him to report anything untoward at the end of his trip. He duly reported afterwards, "I really see nothing to report."[23]

The trip to Zürich came on the heels of his technical assistance mission to Ohio State University. He would not allow his physical ailments to interfere with his activities. However, they were taking their toll. Upon returning from Switzerland he wrote to the Dean of Physical Sciences at Ohio State, "I returned from Switzerland Sunday night. It was an exhausting trip. I am now engaged in finishing up the budget. That is also exhausting."[24]

The Cold War continued to rage throughout 1971. Nevertheless, science began to draw the United States and the U.S.S.R. closer together. Plans were in the works for a September conference of astronomers from America's National Academy of Sciences and the Academy of Sciences of the U.S.S.R. to discuss extraterrestrial intelligence. The two most powerful nations on earth would join forces if it became necessary to do battle with extraterrestrial civilizations.[25]

A less ominous topic was being contemplated for a meeting of mathematicians. In July 1971, Adrian received a letter from the USSR's Academy of Sciences inviting him to attend the International Colloquium on Number Theory in Moscow to be held September 14–18, 1971. The invitation included "some post-colloquium excursions" outside of Moscow. The Soviets graciously offered to defray his expenses while in the U.S.S.R.[26]

Adrian had already made plans to attend his first Executive Committee meeting as Vice President of the IMU; the meeting was scheduled for September 20 in Moscow. The Soviet Academy's invitation to address the Colloquium meshed well with his existing plans and he readily accepted.

The National Security Agency became somewhat alarmed upon learning of the invitation. The Vice-Admiral of the Navy wrote,

> Normally, I would ask that you not make this trip. In this instance, however, there are circumstances which reduce the security risk. Your position in the International Mathematical Union and your preeminence in the field of mathematics add a measure of safety for your visit to Moscow.
> However, your travel to Kiev and Leningrad cause [sic] me more concern. From a security standpoint, I would feel better if you would forego these excursions and limit your itinerary to Moscow....Should any untoward incidents of a security nature occur during your trip, please contact me upon your return.[27]

Adrian reluctantly sent a follow-up letter to the USSR's Academy, demurring that the Alberts' visas required them to depart Moscow on September 20, making "post-colloquium excursions" to Kiev and Leningrad impossible.

On August 30, 1971, the Alberts left Chicago for a month long trip with stops in London, Vienna, Copenhagen, Moscow, and Paris. In Moscow, he delivered a lecture on "Associative Division Algebras" at the Colloquium.[28]

Government minders never left him out of their sight during his stay in Moscow. Upon returning to Chicago at the end of September, he wrote to some friends, "The time...[in Moscow] was really quite unbelievable. I was given the VIP treatment including a student guide and transportation in a private car. It was all very exciting. Some time later we were taken to the Bolshoi Ballet and we really enjoyed it."[29] Frieda remarked afterwards that the ballet was a truly elegant upper crust affair. Despite Russia's purportedly egalitarian society, she had seen no rank and file Russians attending the performance.

Chapter 27

Diplomacy for Mathematics

A.A. Albert used his good offices to expand the boundaries of mathematics and sought to bring more budding scholars within its compass. At the same time, he strove to put the science on a sound financial footing. In short, he was a force for change.

Jacobson characterized him as a "statesman" for the field of mathematics.[1] According to Jacobson, Adrian "had a great deal to do with the establishment of government research grants for mathematics on more or less equal footing with those in other sciences."[2] He did this in each agency where he played a role.

While serving as a member of the General Sciences Panel, an advisory committee to the Defense Department, he fought to reinstate government support for mathematics research at a moment when pure mathematics had once again fallen out of favor in Washington. During a Panel meeting in December 1953, he objected to the "almost complete lack of support for algebra, geometry, and topology...." Afterwards, he reported to the administration at Chicago, "[I] believe I will be successful in restoring the support of pure mathematics which was lost when the 'relevancy' ruling was made."[3]

At the National Science Foundation, he exercised a considerable amount of influence on funding decisions by virtue of his position on the budget committee. In a prepared address to the NSF, he complained that, although mathematics is a key science, it must fight for any support it gets. He deadpanned, "It is still true that in a majority of American universities the way to find the department of mathematics is to ask for the location of the oldest and most decrepit building on campus."[4] In particular, Adrian railed against the notion that only applied, and not pure, mathematics deserved support.

He lobbied for support in the form of postdoctoral fellowships, research projects, transportation to international conferences, and publication costs. As described in an earlier chapter, beginning in the mid-sixties, the Chicago Dean took his message of curriculum improvement on the road.* He visited and/or led teams of consultants to the campuses of American and foreign universities. These site visits culminated in a series of reports with recommendations for revamping mathematics departments and developing new science institutes.

Apart from his experience in curriculum development, there was one more reason for his popularity as a visiting expert. He served as a key consultant to the National Science Foundation, which had earmarked funds for development of science programs. By fiscal year 1970–71, NSF was sitting on appropriations of over $500 million. His stamp of approval on a new research institute, graduate degree program, or mathematics conference was often critical to the NSF's approval.

One of his brainchilds was the idea of holding summer institutes. He pushed the proposal through NSF and selected the first subject and Program Committee. He ran a summer project at Bowdoin College in Maine, several at UCLA, and was connected with one at Stanford.[5]

In 1964 when the Air Force Office of Scientific Research threatened to cut off support for research in pure mathematics, he wrote to the program's director, reminding him, "the whole area of reliable communications and jamproof guidance systems for aircraft and missiles depends fundamentally on algebraic work."[6]

The German algebraist Reinhold Baer, who had worked on group theory, projective planes, and combinatorics in the United States during the forties and fifties, observed that the influential Chicagoan "helped so successfully in raising the economic status of mathematicians."[7]

Adrian was deeply concerned about the short shrift being given to algebra in the nation's schools. During World War II, he described college algebra as "a most abused subject. The time allotted to it is

---

* More about his curriculum development consultancies in Chapter 25.

frequently inadequate for a genuinely good treatment, and indeed the entire course is sometimes omitted."[8]

The rush to introduce calculus into the curriculum as early as possible, he believed, was one of the reasons behind the neglect of algebra on most college campuses. He warned that the practice tends to defeat its own purpose because students cannot understand calculus well without a proper foundation in college algebra. His goal was to correct past abuses by presenting algebra "as a sound and unified whole."[9]

In his intermediate text, *Introduction to Algebraic Theories*, he attempted to make the basic concepts of modern algebra accessible to students whose only previous training in algebra was a beginning course in college algebra. By broadening the base of students with an understanding of modern algebra, he sought to revive what he feared might be a dying science.

Adrian did more than merely complain about the abysmal state of mathematics education in the United States—he took steps to correct it. He was credited with helping the School Mathematics Study Group (SMSG) get off the ground and "helping all through with his wise advice."[10] The SMSG, a group of college and university mathematicians, teachers of mathematics at all levels, experts in education, and representatives of science and technology, began their effort to rewrite mathematics texts in 1958 under the leadership of Edward Begle at Yale University's Mathematics Department. Later, Begle and SMSG moved to Stanford University. In all, the program ran for about 15 years. Begle concluded, "We did what we set out to do; there isn't a textbook being published that hasn't been influenced by SMSG."[11]

Besides expanding the reach of mathematics itself, Adrian sought to cross its boundaries into other sciences. During the late 1930s, he glimpsed the future role of abstract algebra. In one of his early texts, he observed, "During the present century modern abstract algebra has become more and more important as a tool for research not only in other branches of mathematics but even in other sciences."[12]

In a later text, he elaborated on the relevance of algebra, explaining, "During recent years there has been an ever increasing interest in modern algebra not only of students in mathematics but also of those in physics, chemistry, psychology, economics, and statistics."[13]

Adrian reached down to the most elementary level of college mathematics in his book, *College Algebra*. He used the text to reorganize the way algebra was taught in college. The idea was to give students the building blocks essential to establishing an adequate foundation for further study.[14] Continuing his advocacy for expanding the applications of algebra, he predicted that algebra would be useful to social scientists who required a basic understanding of matrices and quadratic forms.[15] He also believed that a proper grounding in college algebra was essential to understanding solid analytic geometry.

Having laid the foundation in his college algebra text for teaching geometry, the Chicago professor completed the process of rewriting basic pedagogical mathematics texts with his book *Solid Analytical Geometry* in 1949. He considered the book as an "extension" of his previous text on college algebra.[16] Professor Albert expressed the hope that "the use of modern algebraic techniques...will serve to make the subject of solid analytic geometry fit better in the teaching of modern mathematics than it has in the past."[17]

It was well known that the University of Chicago offered no curriculum in engineering. The prevailing view at the University was that the school's mission lay in "pure" science. To a minority, Chicago's somewhat elitist perspective stifled innovation.

While Adrian considered himself primarily a pure theoretical mathematician, he believed that applied mathematics was the wave of the future. He successfully advocated in favor of bringing the first computers to campus.

He supported and fostered innovation. With the aid of a grant from IBM, he and some of his colleagues set about offering a program in applied mathematics leading to masters and Ph.D. degrees awarded through the Mathematics Department. To implement the program, he appointed a Committee on Applied Mathematics consisting of five mathematicians (William H. Reid, Felix Browder, Jim Douglas, Jr., I. N. Herstein, and Norman Lebovitz), one professor from the Business School's Committee on Information Sciences, and the brilliant Indian astrophysicist, Subrahmanyan Chandrasekhar, a future Nobel laureate.[18] While the new program did not include as many applied mathematics courses as some Committee members would have liked, it demonstrated that applied mathematics could find a home even among the austere ivory towers at Chicago.

The elder statesman saw the expansion of applied mathematics as a means of offering new opportunities for young mathematicians. At a Youth Conference on the Atom in 1960, he pointed out that, in the

1920s, "the high school graduate bent on a career where he would really use his mathematics actually could only choose an academic career...."[19]

He told the students that such limitations on mathematics as a profession continued until near the end of World War II when "there began to be widespread use for the mathematical model as a means for putting on a sound scientific basis almost every scientific endeavor. Career opportunities began to open up in large numbers in industrial and governmental laboratories and it became possible for a mathematician to use his mathematics in a nonacademic career."[20]

Professor Albert pointed to the development of the high-speed computer as one of the factors increasing the need for mathematically trained personnel at the end of the fifties. He saw the development of high-speed computers as the basis for a new and exciting profession as a numerical analyst.

Citing the shortage of first rate mathematicians in the United States, he encouraged young people attending the Youth Conference to enter the field of mathematics. In the midst of the country's angst over the Soviets' success in launching *Sputnik* into space, he pointed out that mathematics was the one scientific field where the United States still held a clear lead over the other countries of the world.

His conclusions about the country's need for more mathematicians were based upon his work as chairman of the AMS Committee on a Survey of Training and Research Potential in the Mathematical Sciences conducted in the late 1950s, commonly referred to as the "Albert Survey."[21]

Five years after his lecture at the Youth Conference, Dean Albert presented a more optimistic perspective to an audience of scientists at the National Science Foundation. He reported, "there has been a large increase in the number of Ph.D.s during the past ten years. There has also been a dramatic increase in their quality. This nation can now claim to have produced young people who have obtained solutions to some outstanding problems of long standing. This is undoubtedly due in part to the fact that the present prestige of the profession, and the demand for mathematicians in both academic and non-academic pursuits, has attracted some of our most brilliant young men. It is also due to the fact that those who go into the profession are no longer subject to the financial pressures which existed early, and they can now devote themselves to their research. As a result, problems have been tackled and solved which looked hopeless at the start."[22]

He then proceeded to describe some significant advances made by a number of young mathematicians, including those of John Thompson, whose seminal work in group theory had attracted worldwide attention. Adrian's descriptions apparently touched a chord among the scientists in attendance. Afterwards, Milton Rose, head of NSF's Mathematical Sciences Section, wrote to Adrian quoting one of the lecture attendees as "saying, jestfully, that he felt he was a 'purer and better man' for the experience." Rose continued, "Everyone felt that you had conveyed both the excitement of modern mathematical research and its relevance to maintaining, at a national level, a broad scientific strength."[23] The speech achieved its desired effect—getting NSF fired up about supporting research in pure mathematics.

Adrian Albert's views about the role of mathematics challenged the notion of the science as a rather solitary occupation having no direct bearing on the modern world. He knew from his defense work that mathematics could help to break an enemy's secret codes and enable our war planes to hit their targets. Now, he could also see the value of mathematics for peaceful uses in commerce and industry.

Chapter 28

Mathematical Children

A.A. Albert supervised 30 successful Ph.D. candidates, four of whom were women. He became dissertation supervisor for one of them, Marguerite Frank, née Marguerite Straus, by a strange quirk of fate.

Marguerite had an unusual history. Her family fled Paris in 1939 while Hitler was terrorizing Europe. Due to her innate aptitude and the enriched education afforded by a private school in Paris, she had advanced two years beyond her grade level before emigrating. Her family settled in Toronto, Canada, where she attended a rigorous high school that prepared her well for a college major in mathematics.

She went on to study at the University of Toronto, where the 16-year old came under the tutelage of Richard Brauer. She loaded up her undergraduate curriculum with an intensive series of mathematics classes.

At the end of her undergraduate studies, the gifted student earned a fellowship to attend graduate school at Harvard University. At the tender age of twenty, the young woman left family, friends and familiar surroundings in Canada to enter the world of HARVARD.

The terms of her fellowship required the neophyte to teach at Harvard's sister college, Radcliffe. The girl who had just emerged from her teens found the notion of teaching at such a prestigious institution a tad overwhelming.

Although abstract algebra was the specialty that attracted her interest while studying with Brauer, her doctoral supervisor at Harvard would be the country's leading expert in algebraic geometry, Oscar Zariski. Marguerite began to feel like a round peg in a square hole.

In the late 1940s, Harvard had snared Zariski from the University of Illinois with the offer of a chaired professorship. After completing his study of Lefschetz' methods a decade earlier, Zariski proceeded to revolutionize his field by increasingly applying the techniques of modern algebra to algebraic geometry. This work brought him worldwide acclaim. At Harvard, Zariski acquired a god-like stature and gained a reputation as an intimidating figure.[1]

As one of only two women graduate students at Harvard's Mathematics Department, Marguerite felt a heightened pressure to measure up. She was among the best of the best and the competition was stiff. It was not enough to be brilliant when her classmates, as brilliant as she, were competing feverishly in an unrelenting effort to outdo each other. She could not muster up the will to compete head to head with these strivers. The young woman was not yet prepared to focus on mathematics to the exclusion of all else.

A doctoral student's relationship with his or her supervisor may make or break a budding career. The chemistry between Zariski and his student was not going well. She found him to be brilliant, but aloof and inaccessible. It was not so much what he said, but what he did not say. It was not so much his words, but his demeanor. Marguerite sensed a grudging tolerance and disparagement in his glance, and she seemed to freeze in his presence. She completed a Master's Dissertation and received her Master's Degree from Harvard, but felt stymied after three years in Cambridge.

Marguerite's mother noticed her daughter's suffering and agreed to let the struggling student take some time off from the rigors of life at Harvard. The war was over, and Paris now seemed glamorous and deliciously distant from Cambridge, Massachusetts. Marguerite arranged to take classes in history and philosophy in Paris while reassessing her career choices. In Paris, she fell in love. The man was an American with plans to enter a doctoral program at the University of Chicago.

When it came time for the man to leave Paris, Marguerite opted to join him. At Chicago, she found herself at loose ends on campus with no definite plans for her own studies. Although she considered herself a "failure" at mathematics, her affinity for the science remained. She found herself reading several of Emmy Noether's papers on ring theory and jotting down notes about them.

Believing that her work might have some value, she decided to speak to someone in Chicago's Mathematics Department about it. A cursory glance at the University Directory produced the familiar name, A.A. Albert. Marguerite felt sure he would understand both Noether's work and her own. She phoned for an appointment with Professor Albert and he agreed to meet with the timid young woman.

When she arrived in his office, Professor Albert sat at his desk, smoking his pipe. To him, the student with a French accent was a nobody who had dropped in out of the blue. Marguerite introduced herself, but glossed over her previous education. The professor did not press for more information. He glanced at her notes, scarcely looking at her while remaining focused on his pipe. He was not impressed with the work but did not appear dismissive either. "At least he's neutral," she thought, feeling somewhat relieved.

She was struck by the contrast with her former supervisor. He impressed her as approachable, gentle, nonaggressive, and not at all authoritarian. Not wishing to raise expectations by letting him know that Brauer considered her a *wunderkind*, she was content to leave the subject of her past academic accomplishments alone. There was nothing to pique his interest in the student, but he was kind and generous by nature. He handed her a stack of reprints and volunteered, "Well, why don't you take a look at some of my recent papers and come back and see me in a month?"

Marguerite read over the papers and took to them as a duck takes to water. She was intrigued by Professor Albert's work in nonconventional structures. The absentee Harvard student returned in a month feeling a sense of weightlessness and armed with a determination to seek him as her doctoral thesis supervisor.

The pipe-smoking professor appeared rather surprised to learn that the young nobody had fully understood his work. He insisted that she at least register with the University as a "reader" in the Mathematics Department if she wished to work with him. Then he assigned a problem for her dissertation—finding the derivation algebra of one of the noncommutative algebras he had defined.

It was clear he did not take her seriously. One day, Marguerite appeared in his office and announced that her name was no longer Straus, as she had just married a man with the surname, Frank. The professor looked at her in disbelief until she withdrew her hand from the pocket of her raincoat and displayed her wedding ring.

Marguerite dispatched her dissertation problem within a matter of weeks. She had little difficulty in defining the derivation algebra and finding the simple component, a "Lie algebra" of prime characteristic.

These algebras involve a number system whose basis is p, a prime number. They differ from more conventional Lie algebras in that they are finite-dimensional. She wrote out the solution in longhand and brought it to Professor Albert's office.

He took her paper and read it over, then suddenly looked up from his pipe and gazed at the petite personage in front of him as if seeing her for the first time. The student had discovered a brand new class of simple Lie algebras. Prior to her result, only one class of finite-dimensional simple Lie algebras of prime characteristic was known, the class discovered in 1937 by Emmy Noether's former doctoral student, Ernst Witt. Professor Albert could barely contain his excitement about her discovery.

For the first time, he quizzed the young woman about her previous education. Once he learned that she had studied under Brauer and Zariski, her achievement made sense to him. He was so impressed with her findings, he immediately arranged for her work to be published in the prestigious *Proceedings of the National Academy of Sciences*.

On the basis of the work on Lie algebras, he approved her doctoral thesis.[2] Afterwards, Adrian collaborated with Marguerite in developing another new class of finite-dimensional simple Lie algebras over a field of characteristic p and coauthored a paper with her on their new result.[3] The class they discovered later became known as the Albert-Frank-Shalev algebras.

He helped the overachieving visitor find a temporary job working on a game theory research project in Princeton. Eventually, she returned to Harvard, where she took her oral examinations and received a Ph.D. based on the thesis she completed at Chicago.

Her career did not end there. Marguerite discovered yet another of the four known families of simple Lie algebras of prime characteristic and earned high praise for her work on the "Frank-Wolfe algorithm" which was used by the United States Department of Transportation to predict traffic flows. After devoting approximately two decades to responsibilities as a wife and mother interspersed with work on various mathematical research projects, she found a position as a full-time professor. She later credited her doctoral dissertation supervisor with "giving me back some faith in my mathematical ability when I most needed it."[4]

His first female doctoral student, Antoinette Killen Huston, went on to teach at Rensselaer Polytechnical Institute. In later years, she recalled that she did not ask him to take her on as his thesis student. "He came to me and said, 'I have your topic for you.' He offered me my life's fulfillment."[5]

Professor Albert maintained close associations with many of his former doctoral students over the years. The year before stepping down as Dean, he headed a Program Committee planning a "Conference on Non-associative Algebras" at the Illinois Institute of Technology. The 1970 conference, supported by a grant from the National Science Foundation, was the first-ever conference in the United States devoted entirely to nonassociative algebras. Conference planners hoped that the meetings would help to unify existing theories of nonassociative algebras and establish links between these algebras and other areas.

He was ably assisted in planning the program by his former doctoral students, Louis Kokoris and Reuben Sandler, then teaching at the Illinois Institute of Technology and the University of Illinois at Chicago, respectively. Four more of his doctoral students, Richard Schafer of MIT, Richard Block of the University of California, Robert Oehmke of the University of Iowa, and J. Marshall Osborn of the University of Wisconsin lectured at the conference along with algebraists from other universities. Most of them had written their doctoral theses on nonassociative algebras.

The problems that A.A. Albert assigned for doctoral dissertation topics in the 1940s and 1950s influenced the career path of several former Albert doctoral students, who specialized in nonassociative algebras afterwards. They included Schafer, who published a major treatise on nonassociative algebras as well as fifty mathematical papers, mostly dealing with nonassociative algebras. One of Schafer's papers was titled "A Generalization of a Theorem of Albert."[6]

Twenty-three of Adrian's 30 doctoral students became professors at universities and colleges for all or part of their careers. Among those who later chaired their schools' Mathematics Departments were Daniel Zelinsky, Richard Schafer, Eugene C. Paige, J. Marshall Osborn, Robert Oehmke, Murray Gerstenhaber, R. B. Brown, and Richard Block. Several of his students later held sensitive government positions at the National Security Agency and the Aerospace Satellite Navigation Department.

The work that some of these scholars performed as students spawned other important research. The research that Marguerite Frank and others conducted on finite-dimensional simple Lie algebras led to a conjecture by the Russian mathematicians, Kostrikin and Shafarevich, in 1966. The Russian scientists theorized that there were no more than four families of nonclassical simple Lie algebras of prime characteristic, and that these algebras were analogs of algebras studied by Eli Cartan in connection with contact transformations.[7] Block, in collaboration with Robert L. Wilson, later proved their conjecture.[8] Jacobson described the results achieved by Block and Wilson as "marvelous work."[9]

A Ph.D. in mathematics from the University of Chicago opened doors. That, along with a letter of recommendation from Professor Albert, led to prizes, fellowships, grants, professorships, promotions, government posts, and jobs in private industry for Adrian's mathematical offspring. His letters often described the accomplishments of his doctoral students in glowing terms.

The man who had witnessed so much suffering by Ph.D.s unable to find jobs assumed personal responsibility for the welfare of his mathematical children. There was no limit to the amount of effort he was willing to expend in order to help his students find a teaching position. Afterwards, he was often available to answer mathematical questions posed by his protégés.[10]

Before *Sputnik*, the American job market in mathematics was tight and the mathematical community was a relatively small, close-knit group. A well-placed phone call did wonders. Adrian saw his former doctoral student, Eugene Paige, at a summer research project in 1956. At the end of the summer, Paige mentioned that he would like to move from government work to academia. Less than two weeks later, the Chairman of the Mathematics Department at the University of Illinois telephoned Paige to offer him a teaching position.

Marshall Osborn's story is typical of the lengths to which the fatherly professor went in assuring the success of a former doctoral student. Upon Osborn's graduation, Adrian steered him toward a teaching position at the University of Wisconsin where Richard Bruck, a leader in group theory, Osborn's chief interest at the time, served on the faculty. Some years later, when Wisconsin was slow in granting tenure to the deserving professor, the well-connected Chicagoan managed to generate offers for Osborn to teach at two other universities. The tactic succeeded, and Wisconsin granted the tenure.

In November 1970, in honor of my father's sixty-fifth birthday, I decided to turn the tables. After all of the parties at my parents' home, I threw him a surprise party at mine. The party was complete when Kaplansky played "Happy Birthday" on the piano in my living room.

Adrian Albert's term as Dean ended with a flourish. On June 20, 1971, the University of Illinois presented him with the honorary degree of Doctor of Humane Letters, noting, "Dean Albert has long been recognized as one of the world's leading mathematicians, and his writing constitutes a major contribution to the mathematical literature of our century."

Ten days later, he relinquished his post as Dean in accordance with Chicago's mandatory retirement policy. The Director of the Institute for Advanced Study, sensing the bittersweet flavor of the moment, wrote, "I congratulate you on not only surviving but surpassing Deanliness and trust that you find your transition to being entirely free to deal with mathematics and your other interests agreeable."[11]

The University administration presented Adrian with a new appointment letter upon his stepping down from the Deanship. He would return to full-time teaching in the Mathematics Department with a new salary of $40,000, double the peak salary level for mathematics faculty a decade earlier. In addition, he was awarded a three-year "deferred retirement appointment."

Around the time of his retirement as Dean, he received a two-year grant from the National Science Foundation to conduct research on linear algebras. He dove back into his work on pure research with relish.

While stepping down as Dean, he continued to perform many of his former duties. Apart from his research and teaching at the University, he continued his duties as Vice-President of the International Mathematical Union and participated in the boards of the Institute for Advanced Study, the National Research Council's Mathematics Division, and the editorial board of Yeshiva University's journal, *Scripta Mathematica*. He served as a member of the Science and Technology faculty for the Lincoln Academy of Illinois at Baxter Laboratories in Morton Grove, Illinois. And he remained on as consultant for the National Security Agency, the Institute for Defense Analyses, and Argonne National Laboratory.

Since 1969, Adrian had served as "special assistant" to IBM in a program at IBM's research center in Yorktown Heights, New York, where he evaluated cryptographic proposals for data bank security. He continued this consulting function after wrapping up his tenure as Dean.

He became involved in a new undertaking—editing a massive 6-volume collection of L. E. Dickson's mathematical papers. Adrian promised the publishers an expanded version of his previous Dickson memoir as an editor's preface to the collection. Unfortunately, he was unable to keep his commitment.[12]

The Alberts returned from their trip to Russia at the end of September 1971. Robert Brown, Adrian's last doctoral student, had by then joined the faculty at Ohio State University and was co-planning a Conference on Lie Algebras at Ohio State to be held in his former professor's honor.[13] The guest of honor was scheduled to kick off the conference as the opening speaker.

In early October, he wrote to Brown asking that his talk be postponed until the second day of the Conference. The weary traveler explained, "In the week we were in Moscow, I seem to have lost a good deal of weight." It was clear that he was not well.

After returning from Russia in September, his health began a downward spiral. Nonetheless, he remained at his desk through December 1971. Although he was too ill to continue his research, he managed to handle his correspondence. Responding to a National Science Foundation request to rate the work of one of his former doctoral students seeking NSF research support, Adrian commented, "I regard his work as top flight and recommend that it be supported."

His granddaughter recalls that, toward the end of his life, he maintained fond thoughts about his doctoral students. She remembers seeing him chuckle as he affectionately placed a copy of Fanny Boyce's proof regarding nilpotent algebras and idempotents between the pages of one of his texts.[14]

Boyce became the first female mathematics professor at a small Christian school in Illinois, Wheaton College, where she subsequently taught for 32 years.[15] Although Boyce produced no known research papers after receiving her doctorate, her doctoral supervisor thought highly enough of her dissertation on nilpotent algebras to include it in the bibliography of his treatise, *Structure of Algebras*.

Due to Adrian's diabetes, the timing of mealtimes in relation to taking his insulin tablets had always been critical. You could set your watch by dinnertime at the Albert household—it began at precisely six o'clock. Frieda speculated that the visit to Moscow upset the delicate balance between his meals and his medication, and precipitated the sudden decline in his health.

A.A. Albert's fairytale career came to an end when he died on June 6, 1972 of complications from diabetes. He was sixty-six. Only a few months after his demise, *Time* magazine reported the death of his confidante and father figure, Solomon Lefschetz. In a symbolic gesture, Adrian's longtime secretary Dorothy Johnson carefully tucked Lefschetz' obituary into an album of newspaper clippings memorializing Adrian.

Although affected by the past, my father lived his life in the present. I never heard him mention his illustrious ancestor, whom I learned of through other family members. Whether or not that ancestor was the source of his mathematical prowess remains a topic for conjecture. In any case, the inscrutable Elias placed a high value on his son's unusual gift.

Nor did he ever bring up his humble beginnings in Chicago's immigrant neighborhoods. He was not one to dwell upon the obstacles placed in his path. Yet, his incessant drive to over-achieve can be traced to his youth as the son of struggling immigrants—a youth who almost lost the chance to pursue his dream of becoming a mathematician and who never forgot the sacrifices his family made so that he could attend college.

Other factors drove him as well. Behind his mild demeanor, he was a passionate man. Apart from his family, he had two main loves during his lifetime. His first love was mathematics. To Adrian, mathematics was a beautiful, but never fully attainable, mistress. He pursued her with a fervor, day and night. When he left his office, he took her home with him in his mind. He thought about her when he went to sleep at night and when he awoke early in the morning.

His second love was people. He felt for them; he suffered with them. He triumphed when they triumphed. He felt a great deal of pain during the Great Depression for those who received their Ph.D.s in mathematics and had to give up their hopes and dreams because they could not find a job. That empathy shaped his career in many ways. He was constantly searching out means to ensure a reasonable standard of living for his colleagues and for budding mathematicians.

Nothing pleased Adrian more than the success of one of his colleagues. In a series of biographical sketches, mathematicians recounted his reactions to news of a breakthrough by someone he knew. Zelinsky recalled, "one must remember his great, pleased grin that he flashed to welcome news of new successes for any of his extended family anywhere in the world of mathematics."[16] This impression was echoed by his colleague, Izaak Wirszup. "So many times we witnessed his joy at the election of a friend to the National Academy of Sciences or at the elevation of a colleague to a Distinguished Professorship."[17]

It is fair to say that Adrian Albert played a variety of different roles, all of them flowing from his main focus as a researcher of pure mathematics. He recognized that pure research was of no utility to anyone if it sat on a shelf, collecting dust. It was his job as an educator to pass along the results of the research that he and others had conducted in order to allow the field to advance.

Once in the classroom, he became aware that the pedagogical texts then available contained a number of gaps that impeded his students' understanding of the latest developments in mathematics. He undertook a new role, that of author, and wrote a series of texts on modern higher algebra, college algebra, and solid analytic geometry to plug those gaps.

After becoming immersed in the University's culture over a period of years, he saw the need to fulfill still another role—that of administrator. As Chair of the Department of Mathematics, and later, as Dean of Physical Sciences, he sought to fine tune the engine of scientific endeavor at the University to make all of its parts run smoothly.[18] He made every effort to keep faculty members happy so that they could concentrate on their work.

A spin-off of his role as administrator was his role as fund-raiser. Adrian sought out funding from government and industry to provide research monies. He secured funds to underwrite new projects and bring together a synergistic mix of new talent.

The Chicagoan dedicated much of his time to another role—that of United States citizen and patriot. He was not a military man, and never took up arms to defend his homeland. His pen was his sword. He used his knowledge of mathematics during World War II to defend his country through his applied research. Afterwards, he continued his efforts to keep America strong by maintaining its technological edge over other nations.

In his role as good will ambassador, the American mathematician did his best to build bridges of understanding and collegiality with mathematicians all over the world.

As "statesman" for mathematics, he sought to convey an understanding of its value, to prove its relevance, to extend its reach, to bring more young scholars under its umbrella, and to give the science a sound financial footing that would enable it to flourish. As Zelinsky noted, Adrian "was always pleased to use his influence in Washington to improve the status of mathematicians in general...."[19]

In the final decade of his life, he acquired yet another role—what might best be described as "technology transfer agent." The Chicagoan traveled from one university to the next, sharing his expertise in curriculum development and helping to spread seeds of excellence throughout the country.

Notre Dame's tribute recognized the diversity of roles he had played over his career, noting that he was "a mathematician who has given freely of his rich and wide-ranging abilities. At a time when our space program demands increasingly sophisticated mathematical insight, he serves as an important advisor to several government agencies. At a time of educational ferment in all areas, he is active in numerous groups working to improve the quality of mathematics instruction."[20]

Shortly after his death, the International Mathematical Union published a biographical memoir on the organization's newest Vice President. Jacobson was selected to describe his former mathematical sibling and gave this first-hand assessment of Adrian's character: "Albert had wonderful personal qualities: directness, complete honesty without pretense, decisiveness and generosity." He concluded, "These made him very effective in his administrative positions and in his work in IMU."[21]

Perhaps his most meaningful legacy was his impact on mathematicians who continued the work that he had begun. In 1972, the Israeli mathematician S. A. Amitsur presented a paper in which he followed up on a proof developed by Adrian in 1938.[22] Had he known, the American mathematician would have been proud, as he regarded Amitsur as one of the world's leading algebraists.[23]

The following year, the journal *Scripta Mathematica* published a memorial issue dedicated to Adrian's memory.[24] Among the articles in the issue was a paper by I.N. Herstein entitled "On a Theorem of Albert."[25] In it, Herstein gave a new proof for an Albert theorem which, Herstein observed, could be found nowhere in the literature apart from Adrian's book, *Structure of Algebras*.

In a final tribute, Herstein summarized his former colleague's work in mathematics. "What characterized him most as a mathematician was power. This trait of all great mathematicians manifested itself in him right from the start. The problems in which he was interested, and on which he chose to work, were very difficult ones. He attacked them head on, frontally. With an unbelievable tenacity and strength he would hit at them until they succumbed. With this great power of his, and with his insight, he left behind him a monumental legacy in algebra."[26]

A.A. Albert's work on Riemann matrices opened the way for a series of subsequent developments. Gonzalez-Dominguez wrote in 1968 to inquire whether Adrian had read Professor Siegel's lectures. Referring to the work of C. L. Siegel on the theory of Riemann matrices, Gonzalez-Dominguez exclaimed, "Your name appears many times ('an important theorem due to Albert', p. 38; 'Albert has shown', p. 46, 'results in this direction are again due to Albert', p. 50, etc., etc.) The importance Siegel attaches to your results is obvious; it is also obvious that your theorems have been essential for his further work....Your mathematical children are well alive!"[27]

A decade after her husband's demise, Frieda was surprised to receive a letter from a geometer, Louis Auslander, accompanied by an article describing the impact of Adrian's systematic treatment of the theory of the multiplier algebra of a Riemann matrix. He wrote:

> Recently, as a mature mathematician, I became aware of Adrian's work on Riemann matrices and was tremendously impressed by the enormous creative energy that went into his work. The enclosed paper, dedicated to Adrian, was written in an attempt to put Adrian's contribution into a setting that would make it more readily available to the mathematical community. I am certain that future generations will return again and again to study and be impressed by Adrian's contributions.[28]

A few years later, the Mathematics Department at Chicago established the Adrian Albert Memorial Lectures. Each year, three lectures were to be delivered by leading mathematicians. Jacobson arrived from Yale to deliver the first year's lectures, which he devoted to Lie Algebras and Jordan structures—two aspects of nonassociative algebra that had fascinated both men.

The American Mathematical Society embarked on a laborious project to organize and preserve his work. In 1993, the Society republished sixty-six of his most significant papers along with four memoirs in a posthumous collection. Jacobson and four other mathematicians had spent over twenty years assembling the materials, and the AMS presented them in a two volume set, *A. Adrian Albert: Collected Mathematical Papers.*[29]

Frieda continued to reside in the Park Place townhouse for twenty years after her husband's death, drawing solace from the memorabilia that surrounded her. There were photos of him receiving his honorary degrees and small mementos, such as a silver medallion awarded by the Swedish Royal Academy of Science.

Toward the end of her life, I asked whether she regretted not having pursued her own career. She responded, quite simply, "No. I've achieved more in my lifetime than I ever dreamed was possible."

In the spring of 1992, Frieda hosted a bridge party for some of her closest friends. She told her friends about a dream that came to her the previous night in which she saw herself visiting departed loved ones.

The following afternoon, Frieda relaxed in her living room on Park Place, occasionally looking up from the newspaper to savor the sight of cardinals flitting among abundant red berries adorning the firethorn bushes outside her picture window. She finished the crossword puzzle and set the newspaper neatly aside. Then she clicked the remote to watch the five o'clock news and suddenly slipped away. She was eighty-six.

Although A.A. Albert died in 1972, he lives on in the minds of mathematicians and computer experts. In the 1990s, some scientists devised an interactive computer system for building nonassociative algebras.[30] The program enables mathematicians to posit and test hypotheses. They named it "Albert" in his memory.

One of "Albert's" initiators explained that the program was so named because "Albert was the first person to take non-associative algebras seriously and the program is designed to study non-associative algebras."[31] The Albert program, like the man, proved to be efficient. It can accomplish in an hour what previously required several months of effort.[32]

Doctoral Dissertations Written under Supervision of A. Adrian Albert

1934	Antoinette Killen Huston	The integral bases of all quartic fields with a group of order eight.
1934	Oswald Karl Sagen	The integers represented by sets of positive ternary quadratic non-classic forms.
1936	Daniel M. Dribin	Representation of binary forms by sets of ternary forms.
1937	Harriet Rees	Ideals in cubic and certain quartic fields.
1938	Leonard Tornheim	Integral sets of quaternion algebras over a function field.
1938	Sam Perlis	Maximal orders in rational cyclic algebras of composite degree.
1938	Fannie Wilson Boyce	Certain types of nilpotent algebras.
1940	Albert Neuhaus	Products of normal semi-fields.
1941	Frank Lionel Martin	Integral domains in quartic fields.

1941	Anatol Rapoport	Construction of non-Abelian fields with prescribed arithmetic.
1942	Richard D. Schafer	Alternative algebras over an arbitrary field.
1942	Gerhard K. Kalisch	On special Jordan algebras.
1943	Roy Dubisch	Composition of quadratic forms.
1946	Daniel Zelinsky	Integral sets of quasiquaternion algebras.
1950	Charles M. Price	Jordan division algebras and their arithmetics.
1950	Nathan J. Divinsky	Power-associativity and crossed extension algebras.
1951	Murray Gerstenhaber	Rings of derivations.
1951	David M. Merriell	On almost alternative flexible algebras.
1951	Louis Max Weiner	Lie admissible algebras.
1952	John Thomas Moore	Primary central division algebras.
1952	Louis A. Kokoris	New results on power-associative algebras.
1954	Eugene C. Paige	Jordan algebras of characteristic 2.
1954	Marguerite Straus Frank	A new class of simple Lie algebras of characteristic p.
1954	Robert H. Oehmke	A class of non-commutative power-associative algebras.

1956	Richard Earl Block	New simple Lie algebras of prime characteristic.
1957	J. Marshall Osborn	Commutative diassociative loops.
1959	Laurence R. Harper	Some properties of partially stable algebras.
1961	Peter F. G. Stanek	Two-element generation of the symplectic group.
1961	Reuben Sandler	Autotopism groups of some finite nonassociative algebras.
1964	Robert B. Brown	Lie algebras of types E_6 and E_7.

Notes

1: Roots

1. "Jewish Community of Vilna," The Database of Jewish Communities, http://www.bh.org.il/Communities/Archive/Vilna.asp.
2. Frieda D. Albert, typewritten memorandum, undated, Albert family papers.
3. *Jewish Virtual Library, The Vilna Gaon—Rabbi Eliyahu of Vilna (1728–1797)*, from "Great Jewish Leaders," www.us-israel.org/jsource/biography/vilnagaon.html.
4. *Ibid.*
5. Louis Ginzberg, *Students, Scholars and Saints*, (The Jewish Publication Society's Press, 1928), at p. 128.
6. *Encyclopaedia Judaica*, vol. 6, (Keter Publishing House Jerusalem Ltd., 1972), at p. 651.
7. Ginzberg, *supra*, note 5, at p. 130.
8. "Jewish Community of Vilna," *supra*, note 1.
9. Chaim Freedman, *Eliyahu's Branches: The Descendants of the Vilna Gaon and His Family*" (Teaneck, N.J.: Avotaynu, Inc., 1997), at p. 31.
10. Frieda D. Albert memorandum, *supra*, note 2.
11. Frieda D. Albert memorandum, *supra*, note 2.
12. Irving Cutler, *The Jews of Chicago* (Urbana and Chicago: University of Illinois Press, 1996), at p. 57.
13. Interview with Phyllis Cohodes, August, 2003.
14. In his autobiography, mathematician Richard Bellman reports feeling traumatized by the experience of being forced to learn to write with his right hand in grade school. (Richard Bellman, *Eye of the Hurricane: an Autobiography* (Singapore: World Scientific, 1984).)
15. Cutler, *supra*, note 12, at p. 233.
16. Conversation with Aaron Cohodes, September 10, 2004.
17. Travel document from the S. S. Friedriech d Grosse; Frieda D. Albert memorandum, *supra*, note 2.

18. Telephone conversation with Frieda D. Albert, 1992.
19. Audiotaped interview of Frieda D. Albert by Nella Weiner, summer, 1980.
20. Entry by Florence Boyer, January 12, 1922.
21. Entry by Sylvia Kablansky, January 13, 1922.
22. Entry by Bertha Davis, January, 1922.
23. Entry by Ida Fishman, January 14, 1922.
24. Interview with Dan Davis, 2003.

2: Birth of a University in a New City

1. "Chicago: City of the Century," http://pbs.org/wgbh/amex/chicago.
2. Francis Parkman, *The Discovery of the Great West: La Salle*, (New York: Rinehart & Co., 1956), at pp. 44-55.
3. Carl Condit, *The Chicago School of Architecture*, (Chicago: University of Chicago Press, 1964), at p. 16.
4. Thomas Wakefield Goodspeed, *A History of the University of Chicago: The First Quarter-Century* (Chicago: University of Chicago Press, 1916, 1972), at p. 13.
5. *Ibid.*, at pp. 16-17.
6. Thomas Wakefield Goodspeed, *The University of Chicago Biographical Sketches*, vol. 2 (Chicago: University of Chicago Press, 2nd ed., 1925), at pp. 15-19.
7. *Ibid.*, at p. 18.
8. Goodspeed, *supra*, note 4, at p. 19.
9. Goodspeed, *supra*, note 4, at p. 30.
10. Richard Storr, *Harper's University: The Beginnings* (Chicago: University of Chicago Press, 1966).
11. Stanley Appelbaum, *The Chicago World's Fair of 1893*, (New York: Dover Publications, 1980), at p. 1.
12. Daniel Lowe, *Lost Chicago*, (New York: Wings Books, 1975), at p. 10.
13. "Chicago: City of the Century," *supra*, note 1.
14. Carl Condit, *supra*, note 3, at pp. 71-76.
15. Jean Block, *Hyde Park Homes*, (Chicago: University of Chicago Press, 1978), at p. vii; Hyde Park Historical Society, "Paul Cornell—Founder of Hyde Park," http://www.hydeparkhistory.org/cornell.html.
16. Hyde Park Historical Society, *Hyde Park History*, No. 2, 1980.

17. Block, *supra*, note 15.
18. Block, *supra*, note 15, at p. vii.
19. Storr, *supra*, note 10, at pp. 11-12, 17.
20. Articles of Incorporation, in Appendix to Goodspeed, *supra*, note 4, at pp. 478-480.
21. Thomas Wakefield Goodspeed, *The University of Chicago Biographical Sketches*, vol. 1 (Chicago: University of Chicago Press, 2nd ed., 1924), at p. 19.
22. *Ibid.*, at p. 18.
23. Goodspeed, *supra*, note 4, at p. 45.
24. Storr, *supra*, note 10, at p. 44.
25. Lowe, *supra*, note 12, at p. 149.
26. Block, *supra*, note 15, at p. 45.
27. Appelbaum, *supra*, note 11, at p. 3; Condit, *supra*, note 3, at p. 96; Lowe, *supra*, note 12, at pp. 149-150.
28. Block, *supra*, note 15, at p. 50.
29. Http://www.chicago-l.org/figures/yerkes/.
30. Appelbaum, *supra*, note 11, at p. 106, quoting Richard Harding Davis.
31. Block, *supra*, note 15, at p. 53.
32. The poet Harriet Monroe, sister-in-law of the architect, so described the fair in her biography of Root. (David Lowe, *supra*, note 12, at p. 150.)

3. The First Two Generations at the Mathematics Department

1. After Moore's death, Chairman Bliss wrote to Professor H. S. White to verify the details of Moore's role in the 1893 Mathematics Congress. White replied, "Moore was Chairman of the mathematics section of the committee on the Congress of 1893....It was Moore who pushed it and organized the plan and made things go." (Letter to Gilbert Bliss from H. S. White, January 27, 1933, University of Chicago archives).
2. Karen H. Parshall, "Eliakim Hastings Moore and the Founding of a Mathematical Community in America, 1892–1902," in *A Century of Mathematics in America*, Peter Duren et al., eds., vol. 2 (Providence, R.I.: American Mathematical Society, 1988), at p. 163.
3. Ioan James, *Remarkable Mathematicians* (Cambridge University Press; Mathematical Association of America, 2002), at p. 248; Marcus

du Sautoy, *The Music of the Primes* (New York: HarperCollins, 2003), at p. 107.

4. Letter to Gilbert Bliss from H. S. White, January 27, 1933, University of Chicago archives.

5. *Remarkable Mathematicians, supra,* note 3, at pp. 226-227.

6. J.J. O'Connor and E.F. Robertson, "Oskar Bolza," http://www-groups.dcs.st-andrews.ac.uk/history/Mathematicians.

7. J.J. O'Connor and E.F. Robertson, "Leonard Eugene Dickson," http://www-groups. dcs.st-andrews. ac.uk/history/Mathematicians.

8. Departmental list, University of Chicago archives.

9. Letter to Frieda Albert from Oswald Sagen, 1973, Albert family papers.

10. *The Cap and Gown* (Chicago: University of Chicago, vol. 31, 1926), at pp. 38-39.

11. Karen H. Parshall, "Defining a Mathematical Research School: The Case of Algebra at the University of Chicago, 1892–1945," *Historia Mathematica,* vol. 30 (2003).

12. Bibliography of E. H. Moore, University of Chicago archives.

13. Departmental list, *supra,* note 8.

14. Parshall, "Defining a Mathematical Research School," *supra,* note 11.

15. Leonard Eugene Dickson, *Algebras and Their Arithmetics,* (New York: Dover Publications, 1960, repr., 1923 ed.), at p. 220.

16. J. J. O'Connor and E. F. Robertson, "Eliakim Hastings Moore," http://www-groups.dcs.st-andrews.ac.uk/.

17. A. Adrian Albert, "Leonard Eugene Dickson 1874–1954," *Bull. Amer. Math. Soc.* vol. 61 (1955), 331-345, reprinted in *A. Adrian Albert Collected Mathematical Papers,* vol. 1 (Providence, R.I.: American Mathematical Society, 1993), 657-671.

18. J.J. O'Connor and E.F. Robertson, "Gilbert Ames Bliss," http://www-groups.dcs.st-andrews.ac.uk/.

19. Departmental list, *supra,* note 8.

20. W.L. Duren, Jr., "Graduate Student at Chicago in the Twenties," *American Mathematical Monthly,* vol. 83, no. 4 (1976), 243-248, at p. 243.

21. A.A. Albert, typewritten notes, undated, University of Chicago archives.

22. L.E. Dickson, *Algebras and Their Arithmetics* (Chicago: University of Chicago Press, 1923); L. E. Dickson, "New Division Algebras," *Trans. Amer. Math. Soc.,* vol. 28 (1926), 207-234.

23. Duren, *supra,* note 20, at p. 244.

24. Sagen, *supra*, note 9.

25. A.A. Albert, "A Determination of All Associative Algebras in Two, Three, and Four Units Over A Non-Modular Field F," (Master's thesis, University of Chicago, 1927, cited in "Professional Biography of A. Adrian Albert," *A. Adrian Albert Collected Papers*, Block et al., eds.) (Providence, R.I.: American Mathematical Society, 1993)).

26. A.A. Albert, typewritten notes, *supra*, note 21.

27. A.A. Albert, typewritten notes, *supra*, note 21.

28. Daniel Zelinsky, "A. A. Albert," *Amer. Math. Monthly*, vol. 80, no. 6 (June-July, 1973) 661, at p. 662, reprinted in *A. Adrian Albert: Collected Mathematical Papers* (2 vols.) (Providence, R.I.: American Mathematical Society, 1993), at p. xli.

29. Abraham Adrian Albert, "Algebras and Their Radicals, and Division Algebras," (Ph.D. diss., University of Chicago, August, 1928), University of Chicago archives.

30. A. Adrian Albert, *Structure of Algebras* (Providence, R.I.: American Mathematical Society, 1939, 1961 edition), at p. v.

31. A. Adrian Albert, "Leonard Eugene Dickson (1874–1954)," *supra*, note 17. Later, Adrian served as editor of Dickson's collected papers. He was slated to write an Editor's Preface, which would have been an expanded version of his original memoir. However, Adrian passed away before it could be written.

32. Irving Kaplansky, "Abraham Adrian Albert, a Biographical Memoir," *Biographical Memoirs* (National Academy of Sciences, vol. 51, 1980), 3-22.

33. Duren, *supra*, note 20, at p. 244.

34. Duren, *supra*, note 20, at p. 244.

35. A "mentor" is defined as a teacher, tutor, or coach. "Mentorship" is defined as "the influence, guidance, or direction exerted by a mentor." (*Webster's 3rd International Dictionary*, unabridged, 1976).

36. The term "role model" is used here to mean "a person whose behavior in a particular role is imitated by others." (*Merriam-Webster's Collegiate Dictionary*, Tenth Ed., Springfield, Mass., 2002.)

4: The Early Years at Chicago and Princeton

1. Conversation with Rose Mandel, May 7, 1992.

2. Golda Meir, *My Life* (New York: Putnam's Sons, 1975), at p. 24.

3. A. Adrian Albert, "A Determination of All Normal Division Algebras in Sixteen Units," *Trans. Amer. Math. Soc.*, vol. 31 (1929), 253-260.

4. Information provided by Daniel Meyer, Archivist, University of Chicago Library.

5. A A. Albert, Convocation Address, University of Chicago, "The Goal of Excellence in a University," unpub. (August 30, 1963), University of Chicago archives.

6. Information provided by the Historical Society of Princeton.

7. Information provided by Daniel Meyer, Archivist, University of Chicago Library.

8. Albert Tucker interview, Princeton Oral History Project, "The Princeton Mathematics Community in the 1930s," Transcript 29, April 10, 1984, Princeton University Library archives.

9. Adrian's friend and former classmate, Edward ("Jimmy") McShane, described the "slum conditions" at Palmer Library. (Duren, Jacobson, and McShane interview, Princeton Oral History Project, "The Princeton Mathematics Community in the 1930s," Transcript 8, April 10, 1984, Princeton University Library archives.)

10. Nathan Jacobson, "Abraham Adrian Albert, 1905–1972," *Bull. Amer. Math Soc.*, vol. 80 (1974), 1075-1100, at p. 1076.

11. Albert Tucker interview, Princeton Oral History Project, "The Princeton Mathematics Community in the 1930s," Transcript 38, Princeton University Library archives.

12. Jacobson, "Abraham Adrian Albert," *supra*, note 10.

13. J.J. O'Connor and E.F. Robertson, "Solomon Lefschetz," http://www-groups.dcs.st-andrews.ac.uk/.

14. S. Lefschetz, "Reminiscences of a Mathematical Immigrant in the U. S.," Duren et al., eds., *A Century of Mathematics in America*, vol. 1 (Providence, R. I.: American Mathematical Society, 1988), 201-207, at p. 202.

15. J.J. O'Connor and E.F. Robertson, "Solomon Lefschetz," *supra*, note 13.

16. E.J. McShane interview, *supra*, note 9.

17. George W. Brown interview, Princeton Oral History Project, "The Princeton Mathematics Community in the 1930s," Transcript 3, July 25, 1984, Princeton University Library archives.

18. *Time Magazine*, October 16, 1972.

19. Ivan Niven, "The Threadbare Thirties," in *A Century of Mathematics in America*, Peter Duren et al., eds., vol. 1 (Providence, R.I.: American Mathematical Society, 1988), at p. 209.
20. Letter to Dean Henry D. Gale from Gilbert Bliss, November 25, 1930, University of Chicago archives.
21. *Ibid.*
22. *Id.*
23. *Id.*
24. Letter to Professor Lane from A. Adrian Albert, December 22, 1930, University of Chicago archives.
25. *Ibid.*
26. Interview with Alice Schlessinger, Hyde Park Historical Society, March, 2003.
27. Thomas Wakefield Goodspeed, *A History of the University of Chicago, The First Quarter Century* (Chicago: University of Chicago Press, 1916, 1972), at pp. 233-237.
28. *Ibid.*, at pp. 434-436.
29. Frank O'Hara, *The University of Chicago: An Official Guide*, (Chicago: University of Chicago Press, 1928).
30. Eckhart was the largest single contributor to the new mathematics building; however, the majority of the funds was raised from other sources.
31. *Who's Who in Chicago* (A. N. Marquis Company, 1931), at p. 283.
32. Carl Condit, *The Chicago School of Architecture* (Chicago: University of Chicago Press, 1964), at p. 95.
33. *Who's Who in Chicago, supra,* note 31.
34. *Who's Who in Chicago, supra,* note 31.
35. Marshall H. Stone, "Reminiscences of Mathematics at Chicago," *The University of Chicago Magazine* (Autumn 1976), at p. 30.
36. I. N. Herstein, "A. Adrian Albert," *Scripta Mathematica*, vol. 29, no. 3-4 (1973), reprinted in R.E. Block, N. Jacobson, J. M. Osborn, D.J. Saltman and D. Zelinsky, eds., *A. Adrian Albert: Collected Mathematical Papers*, 2 vols, (Providence, R.I.: American Mathematical Society, 1992), lxv-lxvii.
37. A. Adrian Albert, Convocation Address at the University of Chicago, unpub., University of Chicago archives, August 30, 1963.
38. Letter to Helmut Hasse from A. Adrian Albert, February 6, 1934, University of Göttingen archives; Albert Tucker interview, "The Princeton Mathematics Community in the 1930s," transcript 29, April 10, 1984.

39. Story told to N.E. Albert by Frieda Albert; version of Frieda Albert's story recounted to N.E. Albert by J. Glomski, April 10, 2004.

40. Letter to Gilbert Bliss from A. A. Albert, November 18, 1933, University of Chicago archives.

41. Letter to Helmut Hasse from Emmy Noether, Bryn Mawr College, March 6, 1934, quoted in email to N.E. Albert from Peter Roquette, October 17, 2003.

42. *Ibid.*

43. Letter to G. Bliss from A. A. Albert, November 18, 1933, University of Chicago archives.

44. Adrian reported on their meeting in this letter to Chairman Bliss: "Last week R. Brauer of Königsberg, the hypercomplex number man, passed through Princeton on his way to Kentucky where he has a temporary post. Wedderburn, he, and I had a nice long talk." (Letter to G. Bliss from A. A. Albert, November 18, 1933), University of Chicago archives).

45. Brauer later wrote: "I remember also the first time I met Adrian. This was in Princeton in November 1933. I had arrived in America a few days earlier and was still bewildered on my way to Kentucky about which I knew little. When I talked to Adrian, I had immediately the feeling that here was a man with whom I could become a real friend. I believe that Adrian then shared my feeling, and the future has borne this out. We then had very close mathematical interests and we talked a great deal. I also saw Adrian's great scientific enthusiasm, when he took me to see Wedderburn and told him about our conversations." (Letter to "Dear Frieda" from Richard Brauer, June 13, 1972, Albert family papers.)

46. He wrote the following papers during the academic year he spent at the Institute for Advanced Study: A. Adrian Albert: "On Certain Imprimitive Fields of Degree p^2 Over P of Characteristic p, *Ann. of Math.*, vol. 35, no. 2, April, 1934; "Normal Division Algebras Over Algebraic Number Fields," *Bull. Amer. Math. Soc.*, vol. 39 (1933), 746-749; "Normal Division Algebras of Degree 4 over F of Characteristic 2," *Amer. J. Math.*, vol. 56 (1934), 75-86; "Normal Division Algebras Over a Modular Field," *Trans. Amer. Math. Soc.*, vol. 36 (1934), 388-394; "On Normal Kummer Fields Over a Non-Modular Field," *Trans. Amer. Math. Soc.*, vol. 36 (1934), 885-892; "A Solution of the Principal Problem in the Theory of Riemann Matrices," *Ann. of Math.*, no. 2, vol. 34 (1934), 500-515; "Cyclic Fields of Degree p^n over F of Characteristic p," *Bull. Amer. Math. Soc.*, vol. 40 (1934), 625-631; "On

a Certain Algebra of Quantum Mechanics," *Ann. of Math.*, no. 2, vol. 35 (1934), 65-73; and "Integral Domains of Rational Generalized Quaternion Algebras," *Bull. Amer. Math. Soc.*, vol. 40 (1934), 164-176.

47. Albert Tucker Interview, Princeton Oral History Project, "The Princeton Mathematics Community in the 1930s," Princeton University Library archives.

48. J.J. O'Connor and E.F. Robertson, "Oscar Zariski," http://www-groups.dcs.st-andrews.ak/.

49. Letter to Dr. A. Adrian Albert from AAAS, October 5, 1933, University of Chicago archives.

50. Letter to Nathan Jacobson from A.A. Albert, July 11, 1934, University of Chicago archives.

5: Kinships and Kindred Spirits

1. *Cap and Gown*, (University of Chicago, 1926), at p. 269.
2. Correspondence, University of Chicago archives.
3. Interview with Lou Kahn, April 13, 2003.
4. *Ibid.*
5. *Id.*
6. Interview with Richard Schafer, January 10, 2003.
7. Letter to Mrs. Adrian Albert from Richard Block, Riverside, California, June 7, 1972; interview with Richard Block; interview with Robert B. Brown.
8. Interview with E. Kleinfeld, April 23, 2003.
9. Interview with Richard Block, April 19, 2003; interview with Murray Gerstenhaber, April 13, 2003; interview with Daniel Zelinsky, November 4, 2002.
10. Letter to "Dear Frieda" from Tony Huston.
11. Nathan Jacobson, *Collected Mathematical Papers* (Boston: Birkhauser, 1989), at p. 4.
12. *Ibid.*
13. Email to N.E. Albert from Irving Kaplansky, November 7, 2003.
14. Jacobson, *supra*, note 11.
15. Kaplansky recalled that Brauer's arrival at Toronto in 1936 "raised Toronto to a new level of magnitude." (Email to N.E. Albert from Irving Kaplansky, November 10, 2003.)

16. Irving Kaplansky, in G. Benkart, I. Kaplansky, K. McCrimmon, D. Saltman, and G. Seligman, "Nathan Jacobson (1910–1999)," *Notices of the AMS* (October, 2000), at p. 1063.

17. Email to N.E. Albert from Irving Kaplansky, November 7, 2003.

18. *Studies in Modern Algebra*, A.A. Albert, ed., *Studies in Mathematics*, vol. 2 (Englewood Cliffs, N.J.: Mathematical Association of America, Prentice-Hall, distrib., 1963), at p. 1.

19. "The Conference on Algebra at Chicago," *Science*, August 12, 1938, at p. 147.

20. Leo Corry, *Modern Algebra and the Rise of Mathematical Structures* (Basel, Boston, Berlin: Birkhauser Verlag, 1996).

21. Brauer wrote: "It was a few years later, at the time of the first Chicago Algebra Conference that the Brauer-family, Ilse and I, first met the Albert-family, you and Adrian, and for us this was the real beginning of a friendship which has meant a great deal to us ever since. There were the two times we were in Chicago for about half a year, and this would certainly not have been the same without the Alberts." (Letter to "Dear Frieda" from R. Brauer, June 13, 1972, Albert family papers.)

22. Jacobson, *supra*, note 11.

23. Letter to "Dearest" from Adrian, October 31, 1943, Albert family papers.

24. Richard Popp, *Presidents of the University of Chicago* (Chicago: University of Chicago Library, 1992).

6: Mathematese and Mathemagicians

1. E.T. Bell, *Men of Mathematics* (New York: Simon & Schuster, 1965), at pp. 15-16.

2. Lloyd Motz and Jefferson Hane Weaver, *The Story of Mathematics* (New York: Avon Books, 1993), at p. 63.

3. Alfred Hooper, *Makers of Mathematics* (New York: Random House, 1948), at pp. 92-93.

4. Paul J. Nahin, *An Imaginary Tale: The Story of the Square Root of Minus One* (Princeton, N.J.: Princeton University Press, 1998), at p. 54. Nahin credits Norwegian mathematician Caspar Wessel (1745–1818) with discovering the rotational property of the square root of minus one.

5. This example is taken from Walter Fleming and Dale Varberg, *College Algebra* (Englewood Cliffs, N.J.: Prentice-Hall, 1980), at p. 34.
6. Http://www.wikipedia.org/wiki/Associative_algebra.
7. Bell, *supra*, note 1, at p. 352.

7: From Al Jabr to Abstract Algebra

1. Kenneth Travers, Leroy Dalton, and Vincent Brunner, *Using Algebra* (River Forest, Il.: Laidlaw Brothers, A Division of Doubleday, 1945), at p. 1.
2. Alfred Hooper, *Makers of Mathematics* (New York: Random House, 1948), at p. 104.
3. William Karush, *Webster's New World Dictionary of Mathematics* (New York: Webster's New World, 1989), at p. 8.
4. *McGraw-Hill Encyclopedia of Science and Technology*, vol. 1 (New York: McGraw-Hill, 2002), at p. 422.
5. A.A. Albert, *College Algebra* (New York: McGraw-Hill, 1946), at p. 1. This text was updated and revised by Albert in 1963, when it was republished by the University of Chicago Press, Phoenix Science Series.
6. A. Adrian Albert, *Modern Higher Algebra* (Chicago: University of Chicago Press, 1937), at p. vii.
7. A. Adrian Albert, *Fundamental Concepts of Higher Algebra* (Chicago: University of Chicago Press, 1956), at p. v.
8. E.T. Bell, *The Development of Mathematics* (New York: Dover, 1945, 1992), at p. 39. However, J.J. O'Connor and E.F. Robertson report that, in the mid-1930s, a mathematician named Otto Neugebauer managed to decipher the Babylonians' cuneiform writing and learned that Babylonian mathematicians were more advanced than previously believed.
9. Isabella Bashmakova and Galina Smirnova, Abe Shenitzer, trans., *The Beginnings and Evolution of Algebra* (Mathematical Association of America, 2000), at pp. 35-42. While Diophantus introduced symbols for use in algebraic equations in the third century, modern algebraic notation, developed by François Viète (1540–1603), did not arrive until 1591.
10. Rene Descartes, "Rules for Direction of the Mind," *Descartes Selections*, Ralph Eaton, ed. (New York: Charles Scribner's Sons, 1955), 38-83, at p. 54.

11. Bashmakova, *supra*, note 9, at p. 91; E.T. Bell, *Men of Mathematics* (New York: Touchstone—Simon & Schuster, 1937), at p. 47; J. F. Scott, *The Scientific Work of René Descartes* (New York: Garland Publishing, 1987), at p. 88.

12. *Webster's Third International Dictionary* (Springfield, Ma., G. & C. Merriam, 1976).

13. Ivan Grattan-Guiness, *The Norton History of the Mathematical Sciences* (New York: W. W. Norton, 1997), at p. 729.

14. Bell, *Men of Mathematics, supra,* note 11, at pp. 52-53. Although Descartes is widely credited in mathematical literature with inventing the grid system, neither his *La Géométrie*, published in 1637, nor his other published work, provides a description of it.

15. The grid system was most likely developed by Descartes' contemporary, Pierre Fermat (1601–65), who drew on previous work done by ancient Egyptians and Greeks in locating points by means of a system of coordinates. (See J. F. Scott, *The Scientific Work of René Descartes, supra,* note 11, at p. 85.)

16. Dutta, "Cartesian Coordinate System," http://plato.phy.ohiou/edu/.

17. *Ibid.*

18. Motz and Weaver, *The Story of Mathematics* (New York: Avon Books, 1993), at p. 107.

19. J.V. Field, "Girard Desargues," http://www-groups.dcs.st-andrews.ac.uk/.

20. Paul Nahin, *An Imaginary Tale: The Story of the Square Root of Minus One* (Princeton, N.J.: Princeton University Press, 1998), at p.18.

21. Walter Fleming and Dale Varberg, *College Algebra* (Englewood Cliffs, N.J.: Prentice-Hall, 1980), at p. 33.

22. Bashmakova, *supra*, note 9, at p. 117.

23. B.L. van der Waerden, *A History of Algebra* (Berlin: Springer-Verlag, 1985), at p. 76. (Emphasis added.)

24. J.J. O'Connor and E.F. Robertson, "Johann Carl Friedrich Gauss," http://www-groups.dcs.st-andrews.ac.uk/.

25. See, *e.g.*, Calvin Clawson, *The Mathematical Traveler* (Cambridge, Ma.: Perseus Publishing, 2003), at p. 213; E. T. Bell, *supra*, note 11, at p. 233.

26. Nahin, *supra*, note 20, at pp. 80-81.

27. Carl Boyer and Uta Merzbach, *A History of Mathematics* (New York: Wiley & Sons, 1968), at p. 664.

28. Bell, *Men of Mathematics, supra,* note 11, at p. 65.

29. Fleming and Varberg, *supra*, note 21, at p. 14.
30. Eric W. Weisstein, *CRC Concise Encyclopedia of Mathematics* (Boca Raton: Chapman and Hall/CRC, 2003), 2nd ed., at p. 1461.
31. W. Keith Nicholson, *Introduction to Abstract Algebra*, 2d ed. (New York: Wiley & Sons, 1999), at p. 188; J.J. O'Connor and E.F. Robertson, "Julius Wilhelm Richard Dedekind," http://www-groups.dcs.st-andrews.ac.uk/.
32. Conjecture provided by Irving Kaplansky, email to N.E. Albert, March 31, 2003.
33. I.N. Herstein, *Topics in Algebra* (New York: Wiley & Sons, 1975), at p. 207.
34. Michio Kaku, *Hyperspace* (New York: Doubleday Anchor Books, 1994), at pp. 35, 337.
35. "Riemann, (Georg Friedrich) Bernhard," *Encyclopaedia Britannica* (1995), http://www.phy.bg.ac.yu/web_projects/giants/riemann.htm.
36. Stephen Hawking *et al.*, *The Future of Spacetime* (New York: W. W. Norton, 2002), at p. 43; Brian Greene, *The Elegant Universe* (New York: Vintage Books, 1999).
37. *Norton History of Mathematical Sciences*, *supra*, note 13, at pp. 557-565.
38. Arthur Cayley, *Philosophical Transactions*, 1858.
39. Bell, *The Development of Mathematics*, *supra*, note 8, at p. 205.

8: Flowering of Algebra over a Shifting Field

1. I.N. Herstein, *Topics in Algebra*, 2d ed. (New York: John Wiley & Sons, 1975), at p. 1.
2. K.H. Parshall, "Defining a Mathematical Research School: The Case of Algebra at the University of Chicago, 1892–1945," *Historia Mathematica*, vol. 30 (2003).
3. Carl Boyer and Uta Merzbach, *A History of Mathematics* (New York: Wiley & Sons, 1968, 1989), at p. 699.
4. *Ibid.*, at p. 701; Wedderburn, "A Theorem on Finite Algebras," *Trans. Amer. Math. Soc.*, vol. 6 (1905), at pp. 349-352.
5. Charles Curtis, *Pioneers of Representation Theory* (Providence, R.I.: American Mathematical Society, 1999), at p. 201.
6. *McGraw-Hill Dictionary of Mathematics* (New York: McGraw-Hill, 1997), at p. 5.

7. Birkhoff and Mac Lane, *A Survey of Modern Algebra* (Wellesley, Mass.: A. K. Peters, Ltd., 1997), at p. 396.
8. Helmut Hasse, *Higher Algebra*, Theodore Benac, trans., vol. 1 (New York: Frederick Ungar Publishing, 1954), at p. 9.
9. A.A. Albert, "Finite Planes for the High School," *The Mathematics Teacher*, vol. LV, no. 3 (March, 1962).
10. Conversation with Sandra L. Goldberg, November, 2002.
11. Boyer and Merzbach, *supra*, note 3, at pp. 652-653.
12. Harry Henderson, *Modern Mathematics* (New York: Facts on File, Inc., 1996), at p. 51.
13. Parshall, *supra*, note 2.
14. Della Dumbaugh Fenster, "Role Modeling in Mathematics: The Case of Leonard Eugene Dickson (1874–1954)," *Historia Mathematica*, vol. 24 (1997), 7-24, at pp. 11-12.
15. Karen Hunger Parshall, "The Chicago School of Algebra 1892–1950," Abstract, "Mathematical Schools: Italy and the United States at the Turn of the 20th Century," Umberto Bottazzini, University of Palermo, bottazzi@math.unipa.it.
16. Fenster, *supra*, note 14, at p. 19.
17. Http://pages.znet.come/stanleytech/humanities/history/hitler.
18. Constance Reid, *Hilbert* (New York: Copernicus Springer-Verlag, 1996).
19. Constance Reid, *Courant* (New York: Copernicus Springer-Verlag, 1996); Sanford Segal, *Mathematics under the Nazis* (Princeton: Princeton University Press, 2003).
20. Albert Einstein, quoted in Fred Jerome, *The Einstein File* (New York: St. Martin's Griffin, 2002), at p. 24.
21. Ioan James, *Remarkable Mathematicians* (Cambridge, England: Cambridge University Press and Mathematical Association of America, 2002), at p. 256.

9: Explorers in Vector Space

1. A. Adrian Albert, *Modern Higher Algebra* (Chicago: University of Chicago Press, 1937), at p. 217.
2. *Ibid.*; A.A. Albert, *College Algebra* (Chicago: University of Chicago Press, Phoenix Science Series, 1963), at p. 213.

3. See, e.g., Lloyd Motz and Jefferson Hane Weaver, *The Story of Mathematics* (New York: Avon Books, 1993), at p. 251.

4. A.A. Albert, "The Structure of Matrices with any Normal Division Algebra of Multiplications," *Ann. of Math.*, vol. 32 (1931, received by the editors June 16, 1930), 131-148.

5. E.T. Bell, *Men of Mathematics* (New York: Simon & Schuster, 1937, 1965), at p. 360.

6. Interview with Daniel Zelinsky, November 4, 2002.

7. Leonard Eugene Dickson, *Algebras and Their Arithmetics* (New York: Dover Publications, 1960, repub., 1923 ed.), at p. 59; Daniel Zelinsky, "A. A. Albert," *Amer. Math. Monthly*, vol. 80, no. 6 (June-July 1973), 661-665, at p. 663.

8. Irving Kaplansky, "Abraham Adrian Albert, A Biographical Memoir," *Biographical Memoirs* (National Academy of Sciences, vol. 51, 1980), 3-22, at pp. 4-5.

9. Email to N.E. Albert from D. Zelinsky, February 12, 2003.

10. Nathan Jacobson, "Abraham Adrian Albert, 1905–1972," *Bull. Amer. Math. Soc.*, vol. 80, no. 6, (1974), 1075-1100, at p. 1079.

11. W. Keith Nicholson, *Introduction to Abstract Algebra* (New York: Wiley, 1999), at p. 542.

12. L.E. Dickson, "New Division Algebras," in *Dickson Collected Papers*, A. A. Albert, ed., at p. 207.

13. Albert, *Modern Higher Algebra, supra*, note 1, at p. 242.

14. "Albert, Abraham Adrian," *McGraw-Hill Modern Men of Science* (New York: McGraw-Hill, 1966), at p. 5; Jacobson, *supra*, note 10, at p. 1079.

15. A. Adrian Albert, "On the Structure of Normal Division Algebras, *Ann. of Math.*, vol. 30, no. 2 (1929), 322-338, at p. 322.

16. Interview with Erwin Kleinfeld, April 23, 2003.

17. A. Adrian Albert, "On the Structure of Normal Division Algebras," *Ann. of Math.*, vol. 2 (1929), 322-338, at p. 322.

18. A. Adrian Albert, "Algebras and Their Radicals, and Division Algebras," Ph.D. diss., August, 1928, University of Chicago archives; Zelinsky, *supra*, note 7, at p. 663.

19. A. Adrian Albert, "A Determination of All Normal Division Algebras in Sixteen Units," *Trans. Amer. Math. Soc.*, vol. 31 (1929), 253-260; Kaplansky, *supra*, note 8, at p. 5; Jacobson, *supra*, note 10, at p. 1079.

20. Kaplansky, *supra*, note 8.

21. Zelinsky, "A. A. Albert," *supra*, note 7, at p. 663.

22. Birkhoff and Mac Lane, *A Survey of Modern Algebra* (Wellesley, Mass.: A.K. Peters, Ltd., 1997), at p. 414; Richard A. Dean, *Elements of Abstract Algebra* (New York: Wiley & Sons, 1966), at p. 203.

23. Dickson, *supra*, note 7, at p. 59.

24. A. Adrian Albert, "New Results in the Theory of Normal Division Algebras," *Trans. Amer. Math. Soc.*, vol. 32 (1930), 171-195.

25. *Ibid.*, at p. 188.

10: The Proof

1. L.E. Dickson, *Algebren und Ihre Zahlentheorie* (Zurich: Orell Fussli, 1927).

2. Charles W. Curtis, *Pioneers of Representation Theory* (Providence, R.I.: American Mathematical Society, 1999.)

3. Peter Roquette, "The Brauer-Hasse-Noether Theorem in Historical Perspective," 2/28/04 version, http://www.roquette.uni-hd.de.

4. *Ibid.*

5. *Id.*

6. Letter to Helmut Hasse from Emmy Noether, December 19, 1930, quoted in Roquette, *supra*, note 3; A. Adrian Albert, "New Results in the Theory of Normal Division Algebras," *Trans. Amer. Math. Soc.*, vol. 32 (1930), 171-195.

7. Email to N.E. Albert from P. Roquette, October 14, 2003.

8. Email to N.E. Albert from P. Roquette, December 7, 2003.

9. Letter to H. Hasse from A. A. Albert, March 23, 1931, University of Göttingen archives.

10. D. Fenster characterized the cordial letters written by the 25-year old Adrian Albert to Helmut Hasse as "a model of diplomacy" reflective of a maturity beyond his years. (Della Fenster, "Leonard Dickson's Algebraic Research and Its Institutional Context: A Case Study of A. Adrian Albert," lecture presented at M.S.R.I., San Francisco, Ca., April, 2003.)

11. A. Adrian Albert: (1) "A Note on an Important Theorem on Normal Division Algebras," *Bull. Amer. Math. Soc.*, vol. 36 (1930), 649-650; (2) "New Results in the Theory of Normal Division Algebras," *Trans. Amer. Math. Soc.*, vol. 32 (1930) (received by the AMS in December, 1929), 171-195; (3) "On Direct Products," *Trans. Amer. Math. Soc.*, vol. 33 (June, 1931) (received by the AMS on April 17, 1931), 690-

711; (4) "Normal Division Algebras of Degree Four Over an Algebraic Field," *Trans. Amer. Math. Soc.*, vol. 34 (1932) 444-456 (received by the AMS on September 29, 1931 and previously reported in the June, 1931 *Proceedings of the National Academy of Sciences*); and (5) "Division Algebras Over an Algebraic Field," *Bull. Amer. Math. Soc.*, vol. 37 (October, 1931) (presented to the AMS on September 9, 1931), 777-784.

12. Letters to H. Hasse from A.A. Albert, March 23 and 27, 1931, University of Göttingen archives.

13. Roquette, *supra*, note 3; I. Kaplansky, "Abraham Adrian Albert, A Biographical Memoir," *Biographical Memoirs* (National Academy of Sciences, vol. 51, 1980), 3-22, at p. 5.

14. A.A. Albert, "On Direct Products," *supra*, note 11; N. Jacobson, "Abraham Adrian Albert, 1905–1972," *Bull. Amer. Math. Soc.*, vol. 80, no. 6, (1974), 1075-1100, at p. 1081.

15. Letter to H. Hasse from A.A. Albert, May 11, 1931, University of Göttingen archives.

16. *Ibid*.

17. A. Adrian Albert, "Normal Division Algebras of Degree Four Over an Algebraic Field," *supra*, note 11.

18. Letter to Hasse from Albert, *supra*, note 15.

19. Adrian referred to Hasse's request in a June 30, 1931 reply to Hasse.

20. A. Adrian Albert, "Normal Division Algebras of Order 2^{2m}," *Proc. Nat. Acad. Sci. U.S.A.*, vol. 17 (1931), 389-392.

21. Letter to H. Hasse from A.A. Albert, June 30, 1931, University of Göttingen archives.

22. *Ibid*.

23. A. Adrian Albert, "Division Algebras Over an Algebraic Field," *supra*, note 11; N. Jacobson, *supra*, note 14, at p. 1081.

24. A. Adrian Albert and Helmut Hasse, "A Determination of All Normal Division Algebras over an Algebraic Number Field," *Trans. Amer. Math. Soc.*, vol. 34 (1932), 722-726, at p. 723.

25. Jacobson, *supra*, note 14, at p. 1080.

26. Roquette, *supra*, note 3.

27. Letter to H. Hasse from A.A. Albert, November 4 or 6, 1931. (The typewritten date of Albert's letter is November 4, but it is overwritten in ink as November 6.)

28. Albert and Hasse, *supra*, note 24, at p. 723.

29. Letter to Hasse from Albert, November 4 or 6, 1931, *supra*, note 27.

30. Roquette, *supra*, note 3, at p. 7; Albert and Hasse, *supra*, note 24, at p. 724.

31. R. Brauer. H. Hasse, and E. Noether, "Beweis eines Hauptsatzes in der Theorie der Algebren," ("Proof of a Main Theorem in the Theory of Algebras"), *Journal für die Reine und Angewandte Mathematik* (Journal for Pure and Applied Mathematics, also known as *Journal für Mathematik* and commonly called *Crelle's Journal* after its founder), vol. 167 (1932), 399-404.

32. Letter to H. Hasse from A.A. Albert, November 26, 1931, University of Göttingen archives.

33. *Ibid.*

34. *Id.*

35. R. Brauer, H. Hasse, and E. Noether, *supra*, note 31, translated by P. Roquette in email to N. E. Albert, October 14, 2003.

36. Letter to H. Hasse from A.A. Albert, December 9, 1931, University of Göttingen archives.

37. Jacobson, *supra*, note 14, at p. 1081; interview with D. Zelinsky, November 4, 2002.

38. Letter to H. Hasse from A.A. Albert, January 25, 1932, University of Göttingen archives.

39. Letter to H. Hasse from A.A. Albert, June 22, 1932, University of Göttingen archives. In the letter, Albert tells Hasse, "Your proof sheets of our joint Transactions paper reached me and I was very pleased with essentially all of your revisions.... I sent your proof sheets to...New York..."

40. Albert and Hasse, *supra*, note 24, at p.724.

41. Email to N.E. Albert from Louis Rowen, May 1, 2003.

42. Roquette, *supra*, note 3.

43. Roquette, *supra*, note 3.

44. Letter to H. Hasse from A.A. Albert, *supra*, n. 38.

45. Roquette, *supra*, note 3.

46. Letter to H. Hasse from E. Artin, quoted in Roquette, *supra*, note 3.

47. A. Adrian Albert, *Structure of Algebras* (New York: American Mathematical Society Colloquium Publications, vol. 24, 1939), at p. v. (revised printing, Providence, R.I.: American Mathematical Society, 1961).

48. Jacobson, *supra*, note 14, at page 1081.

49. Email to N.E. Albert from D. Zelinsky, September 19, 2003.

50. Email to N.E. Albert from P. Roquette.

51. Jacobson, *supra*, note 14, at page 1082; A. Adrian Albert: 1) "On Certain Imprimitive Fields of Degree p^2 of Characteristic p," *Ann. of Math.*, vol. 35, no. 2 (1934), 211-219; 2) "Cyclic Fields of Degree p^n Over F of Characteristic p," *Bull. Amer. Math. Soc.*, vol. 40 (1934), 625-631; 3) "Normal Division Algebras of Degree p^e over F of Characteristic p," *Trans. Amer. Math. Soc.*, vol. 39, 1936, 183-188; 4) "Simple Algebras of Degree p^e over A Centrum of Characteristic p," *Trans. Amer. Math. Soc.*, vol. 40 (1936), 112-126; 5) "p-Algebras Over a Field Generated by One Indeterminate," *Bull. Amer. Math. Soc.*, vol. 43 (1937), 733-736; 6) "On Cyclic Algebras," *Ann. of Math.*, no. 2, vol. 39 (1938).

52. A. Adrian Albert, "On the Construction of Cyclic Algebras with a Given Exponent," *Amer. J. Math.*, vol. 54 (January, 1932), 1-13, at p. 9.

53. Letter to Helmut Hasse from A. Adrian Albert, April 1, 1932, University of Göttingen archives.

54. *Ibid.*

55. A. Adrian Albert, *Structure of Algebras*, *supra*, note 47, at p. 109.

56. J. Minac and A. Wadsworth, "The First Two Cohomology Groups of Some Galois Groups," http://www.math.ucsd.edu/wadswrth/MW.html.

57. A.A. Albert, "A Consideration of Noncyclic Normal Division Algebras," *Bull. Amer. Math. Soc.*, vol. 38 (1932), 449-456.

58. Constance Reid, *Courant* (New York: Springer-Verlag, 1996), at p. 157.

59. Marcus du Sautoy, *The Music of the Primes* (New York: HarperCollins, 2003) at p. 251.

60. Sanford Segal, *Mathematicians under the Nazis* (Princeton: Princeton University Press, 2003), at p. 159.

61. See report of interview of Hasse in Reid, *supra*, note 58, at p. 250.

62. Segal, *supra*, note 60, at p. 159.

63. A.A. Albert, "On Normal Kummer Fields over a Non-modular Field," *Trans. Amer. Math. Soc.*, vol. 36 (1934), 885-892.

64. Handwritten draft of letter to A.A. Albert from H. Hasse, January 22–February 2, 1935, University of Göttingen archives.

65. Email to N.E. Albert from P. Roquette, June 2, 2004, quoting letter to H. Hasse from C.L. Siegel, December 13, 1935.

66. Email to N.E. Albert from P. Roquette, June 2, 2004, quoting letter to C.L. Siegel from H. Hasse, December 19, 1935.

67. A.A. Albert, "Non-cyclic Algebras with Pure Maximal Subfields," *Bull. Amer. Math. Soc.*, vol. 44 (1938), 576-579.

68. Marshall Stone's extract of letter to Stone from Helmut Hasse, March 25, 1939 (incorporated into Stone's reply), AMS archives, Box 27, Folder 59.

69. *Ibid.*

70. Letter to Helmut Hasse from Marshall Stone, May 2, 1939, AMS archives, Box 27, Folder 59.

71. J.J. O'Connor and E.F. Robertson, "Helmut Hasse," http://www-groups.dcs.st-andrews.ac.uk/

11: Traveling Riemann's Universe

1. Lloyd Motz and J. H. Weaver, *The Story of Mathematics* (New York: Avon Books, 1993), at pp. 233-234.

2. Michio Kaku, *Hyperspace* (New York: Anchor Books, 1994), at p. 36.

3. L. Auslander and R. Tolmieri, "A Matrix Free Treatment of the Problem of Riemann Matrices," *Bull. Amer. Math. Soc.*, vol. 5, no. 3 (1981), 263-312.

4. *Ibid.*, at p. 269.

5. Irving Kaplansky, "Abraham Adrian Albert, A Biographical Memoir," *Biographical Memoirs* (National Academy of Sciences, vol. 51, 1980), 3-22, at p. 6.

6. A. Adrian Albert: (1) "On the Structure of Pure Riemann Matrices with Non-commutative Multiplication Algebras," *Proc. Nat. Acad. Sci.*, vol. 16 (1930), 308-312; (2) "On Direct Products, Cyclic Division Algebras, and Pure Riemann Matrices," *Proc. Nat. Acad. Sci.*, vol. 16 (1930), 313-315; (3) "On Direct Products, Cyclic Division Algebras, and Pure Riemann Matrices," *Trans. Amer. Math. Soc.*, vol. 33 (1931), 219-234; (4) "The Non-existence of Pure Riemann Matrices with Normal Multiplication Algebras of Order Sixteen," *Ann. of Math.*, vol. 31, no. 2 (1930), 375-380; and (5) "The Structure of Pure Riemann Matrices with Non-commutative Multiplication Algebras," *Palermo Rendiconti*, vol. 55 (1931), 57-115.

7. Daniel Zelinsky, "A.A. Albert," *Amer. Math. Monthly*, vol. 80, no. 6 (June–July 1973), 661-665, at p. 664.

8. N. Jacobson, "Abraham Adrian Albert, 1905–1972," *Bull. Amer. Math. Soc.*, vol. 80, no. 6, (1974), 1075-1100, at p. 1084 (emphasis added); A. A. Albert: (5) "On the Construction of Riemann Matrices I," *Ann. of Math.*, vol. 35, no. 2 (1934), 1-28; (6) "A Solution of the Principal

Problem in the Theory of Riemann Matrices," *Ann. of Math.*, vol. 35, no. 2 (1934), 500-515; (7) "On the Construction of Riemann Matrices II," *Ann. of Math.*, vol. 36, no. 2 (1935), 376-394.

9. Letter to N.E. Albert from George Glauberman, September 17, 2003; email to N.E. Albert from Louis Rowen, May 1, 2003.

10. A. Adrian Albert, "Involutorial Simple Algebras and Real Riemann Matrices," *Ann. of Math.*, vol. 36 (1935), 886-964; A. Adrian Albert, *Structure of Algebras* (New York: *American Mathematical Society Colloquium Publications*, vol. 24, 1939), at p. v. (revised printing, Providence, R.I.: American Mathematical Society, 1961), at pp. 151-170.

11. Email to N.E. Albert from Louis Rowen, May 1, 2003.

12. Marcus du Sautoy, *The Music of the Primes* (New York: Harper Collins, 2003), at p. 224; J.J. O'Connor and E.F. Robertson, "Frank Nelson Cole," http://www-groups.dcs.st-andrews.ac.uk/.

13. A. Adrian Albert, *Structure of Algebras, supra,* note 10, at p. vi.

14. Max-Albert Knus, Alexander Merkurjev, Marcus Rost, and Jean-Pierre Tignol, *The Book of Involutions* (Providence, R.I.: American Mathematical Society, Colloquium Publications), vol. 44, 1998, at p. 68.

15. *Ibid.*, at p. 67.

16. *Id.*, at p. 68.

17. Constance Reid, *Courant* (New York: Springer-Verlag, 1996), at p. 156.

18. *Deutsche Mathematik*, vol. 1 (1936), 197-238.

19. A. Adrian Albert, "On the Construction of Cyclic Algebras with a Given Exponent," *Amer. J. Math.*, vol. 54 (1932), 1-13.

20. Email to D. Zelinsky from P. Roquette, November 27, 2003.

21. A. Adrian Albert, *Modern Higher Algebra* (Chicago: University of Chicago Science Series, University of Chicago Press, 1937).

22. The textbook remains in print today and, in the opinion of one reviewer, "on certain subjects it is an indispensable reference." (Kaplansky, *supra*, note 5, at p. 6.)

23. Letter to Prof. A.A. Albert from R.G.D. Richardson, Secretary, American Mathematical Society, September 12, 1939; letter to Prof. A.A. Albert from J.R. Kline, Secretary, American Mathematical Society, January 4, 1943, University of Chicago archives.

24. Letter to Prof. A.A. Albert from R.G.D. Richardson, Secretary, American Mathematical Society, September 14, 1937, University of Chicago archives.

25. Letter to Frieda from Adrian, Madison, Wisconsin, September 5, 1939, Albert family papers.

26. Jacobson, *supra*, note 8, at p. 1077.

27. A. Adrian Albert, *Structure of Algebras, supra*, note 10.

28. Irving Kaplansky later described *Structure of Algebras* as a definitive treatise on algebras that remains a classic. (Kaplansky, "Abraham Adrian Albert," *supra*, note 5, at p. 6.) In 1973, I.N. Herstein referred to a "beautiful theorem" that Adrian developed in his work on Riemann matrices, and observed, "the only place, as far as I know, where it appears in the literature is in Albert's book." (I.N. Herstein, "On a Theorem of Albert," *Scripta Mathematica*, vol. 29, no. 3-4 (1973), at p. 391.) Over 30 years after its publication, Jacobson observed, "This extremely readable and beautifully organized book can still be recommended to a beginning student with a serious interest in structure theory and is an indispensable reference book for certain aspects of the theory, particularly in the theory of p-algebras, and of algebras with involution." (Jacobson, *supra*, note 8, at p. 1082.)

29. Letter to J.R. Kline, Secretary, from Tomlinson Fort, June 20, 1942, AMS archives, Box 27, Folder 132.

30. Letter to Tomlinson Fort from J.R. Kline, June 22, 1942, AMS archives, Box 27, Folder 132.

31. Carbon copy of letter to Murnaghan from A.A. Albert, August 14, 1942 with handwritten note to J.R. Kline, AMS archives, Box 28, Folder 45.

32. Letter to E.J. McShane and R.L. Wilder from A.A. Albert, July 16, 1945, AMS archives, Box 30, Folder 118; Letter to A.A. Albert from J.R. Kline, Secretary, August 10, 1945, AMS archives, Box 30, Folder 118; Letter to J.R. Kline from A.A. Albert, May 1, 1944, AMS archives, Box 29, Folder 15.

33. Karl Sabbagh, *The Riemann Hypothesis: The Greatest Unsolved Problems in Mathematics* (New York: Farrar, Straus and Giroux, 2002), at pp. 108-109.

34. "Riemann Hypothesis," Clay Mathematics Institute, http://www.claymath.org/Millennium_Prize_Problems.

35. Sabbagh, *supra*, note 33, at p. 109, quoting *Time* magazine.

36. Letter to Professor O. Veblen, Institute for Advanced Study School of Mathematics from A. Adrian Albert, December 13, 1939, University of Chicago archives.

37. Letter to Adrian from John von Neumann, November 14, 1940; letter to Professor John von Neumann from Adrian Albert, December 27, 1940, University of Chicago archives.

38. Letter to Professor John von Neumann from Adrian Albert, March 3, 1941, University of Chicago archives.
39. Letter to Dr. Robert Oppenheimer from Marston Morse, December 15, 1954, Princeton University archives.
40. Email to N.E. Albert from J. Marshall Osborn, June 13, 2003.

12: A New School of Algebra

1. Constance Reid, *Hilbert* (New York: Springer-Verlag, 1996), at p. 104.
2. J.J. O'Connor and E.F. Robertson, "Quotations by Hermann Weyl," http://www-groups.dcs.st-andrews.ac.uk/.
3. Nathan Jacobson, "Abraham Adrian Albert, 1905–1972," *Bull. Amer. Math. Soc.*, vol. 80 (1974), 1075-1100 at p. 1076.
4. Michael Bass, "Quantum Mechanics: The Other Great Revolution of the 20th Century—Part III," http://216.239.53.104/search?q=cache:VIHkqpeuKIgJ:caos.creol.ucf.edu/.
5. P. Jordan, J. von Neumann and E. Wigner, "On An Algebraic Generalization of the Quantum Mechanical Formalism," *Ann. of Math.*, vol. 35 (1934), 29-64; Nathan Jacobson, *supra*, note 3, at p. 1086.
6. Jacobson, *supra*, note 3, at p. 1086; Holger P. Petersson, "Albert Algebras: Their Meaning for Present-Day Mathematics," unpub., in email to N.E. Albert, June 10, 2003.
7. A. Adrian Albert, "On a Certain Algebra of Quantum Mechanics," *Ann. of Math.*, vol. 35, no. 1 (1934), 65-73.
8. Kevin McCrimmon, *A Taste of Jordan Algebras* (New York: Springer-Verlag, 2003).
9. A.A. Albert, "Non-associative algebras. I. Fundamental Concepts and Isotopy," *Ann. of Math.*, vol. 43, no. 4 (1942), 685-707; A.A. Albert, "Non-associative algebras. II. New Simple Algebras," *Ann. of Math.*, vol. 43, no. 4 (1942), 708-723.
10. Irving Kaplansky, "Abraham Adrian Albert, A Biographical Memoir," *Biographical Memoirs*, National Academy of Sciences, vol. 51 (1980), 3-22, at p. 8.
11. A.A. Albert, "Non-Associative Algebras I," *supra*, note 9, at p. 685. (Emphasis added.)
12. Interview with Daniel Zelinsky, May 7, 2003.
13. A.A. Albert, "Nonassociative Algebras I," *supra*, note 9, at p. 696.
14. Jacobson, *supra*, note 3, at p.1092.

15. Jacobson, *supra*, note 3, at p. 1090.

16. Jacobson, *supra*, note 3, at p. 1090; A.A. Albert, "The Radical of a Non-Associative Algebra," *Bull. Amer. Math. Soc.*, vol. 48 (1942), 708-723.

17. A.A. Albert, "On Nonassociative Division Algebras," *Trans. Amer. Math. Soc.*, vol. 72 (1952), 296-309, at p. 306.

18. A.A. Albert, "Finite Division Algebras and Finite Planes," *Proc. Sympos. Appl. Math.*, vol. 10, Amer. Math. Soc. (1960), 53-70, at p. 69; A. A. Albert, "Generalized Twisted Fields," *Pacific J. Math.*, vol. 11 (1961), 1-8.

19. A.A. Albert, "On the Collineation Groups Associated with Twisted Fields," *Calcutta Math. Soc., Golden Jubilee Commemoration Volume*, 1958–59, part II, Calcutta Math. Soc. (1963), 485-497; A.A. Albert, "Generalized Twisted Fields," *ibid.*; A.A. Albert, "Isotopy for Generalized Twisted Fields," *An. Acad. Brazil. Ci.*, vol. 33 (1961), 265-275.

20. A.A. Albert, "On Jordan Algebras of Linear Transformations," *Trans. Amer. Math. Soc.*, vol. 59 (1946), 524-555, at p. 524; Nathan Jacobson, "Abraham Adrian Albert (1905–1972)," *IMU Bulletin of the International Mathematical Union*, No. 4 (December 1972), at p. 9.

21. Kaplansky, *supra*, note 10, at p. 7.

22. A.A. Albert: 1) "On Jordan Algebras of Linear Transformation," *Trans. Amer. Math. Soc.*, vol. 59 (1946), 524-555; 2) "A Structure Theory for Jordan Algebras," *Ann. of Math.*, vol. 48 (1947), 546-567; 3) "A Theory of Power-Associative Commutative Algebras," *Trans. Amer. Math. Soc.*, vol. 69 (1950), 503-527.

23. Kaplansky, *supra*, note 10, at p. 7.

24. Interview with J. Marshall Osborn, April 6, 2003.

25. A. Adrian Albert, "Quasigroups I," *Trans. Amer. Math. Soc.*, vol. 54 (1943), 507-519; A. Adrian Albert, "Quasigroups II," *Trans. Amer. Math. Soc.*, vol. 55 (1944), 401-419.

26. Interviews with J.M. Osborn, April 6 and April 9, 2003; interview with Richard Block, April 19, 2003.

27. A.A. Albert, "On the Power-associativity of Rings," *Summa Brasil. Math.*, vol. 2, no. 2 (1948), 21-33; Jacobson, *Bull. Amer. Math. Soc.*, *supra*, note 3, at p. 1089.

28. A.A. Albert, "Power-Associative Rings," *Trans. Amer. Math. Soc.*, vol. 64 (1948), 552-593.

29. J.V. Kadeisvili, "Open Research Problems in Mathematics," ch. V, "Open Research Problems in Lie-Santilli Theory," http://www.magnegas.com/science/ir00018.htm.

30. Letter to Dear Sirs from A.T. Gainov, Mathematical Institute, Siberian Division, Academy of Sciences of the USSR, November 21, 1972.

31. A.A. Albert, "On Right Alternative Algebras," *Ann. of Math.*, vol. 50, no. 2 (1949), 318-328.

32. A. Adrian Albert, "Almost Alternative Algebras," *Portugaliae Mathematica*, vol. 8, Fasc. 1 (1949), 23-36.

33. A.A. Albert, "On Simple Alternative Rings," *Canad. J. Math.*, vol. 4 (1952), 129-135; Interviews with Erwin and Margaret Kleinfeld, April 23, 2003.

34. Kaplansky, *supra*, note 10, at p. 9. Kleinfeld reports that A.A. Albert put together a small 1951 conference on alternative rings designed to share information among those who had innovated in the area. Attendees included Max Zorn, Richard Bruck, Irving Kaplansky, Erwin Kleinfeld, Hans Zassenhaus, and A.A. Albert, among others. (Interview with Erwin Kleinfeld, April 23, 2003.)

35. Interview with R. Schafer, January 9, 2003; interview with George Glauberman, January 15, 2003.

36. Jacobson, *Bull. Amer. Math. Soc.*, *supra*, note 3, at p. 1090; A.A. Albert, unpub. report of current activities, December 1954, University of Chicago archives.

37. A.A. Albert and M.S. Frank, "Simple Lie Algebras of Characteristic p," *Univ. e Politec. Torino. Rend. Sem. Mat.*, vol. 14 (1954–55), 117-139; Interview with R. Block, April 19, 2003.

38. Nathan Jacobson interview, Princeton Oral History Project, "The Princeton Mathematics Community in the 1930s," Transcript 8, April 10, 1984, Princeton University Library archives.

39. Interview with Daniel Zelinsky, November 4, 2002.

40. *McGraw-Hill Modern Men of Science* (New York: McGraw-Hill, 1966), at p. 6.

41. Kaplansky, *supra*, note 10, at p. 8.

42. D. Zelinsky, "A.A. Albert," *Amer. Math. Monthly*, vol. 80, no. 6 (1973), 661-665, at p. 664.

43. *Ibid.*, at p. 664.

44. Interviews with Daniel Zelinsky, November 4 and November 6, 2002.

45. Walter Jacobs, Computation Division, U.S. Air Force, "Book Reviews," *Science*, vol. 124, December 28, 1956, at p. 1296.
46. Jacobson, *Bull. Amer. Math. Soc., supra*, note 3, at p. 1093.
47. Jacobson, *IMU Bulletin, supra*, note 20, at p. 10.

13: Mathematics during World War II

1. Letter to Helmut Hasse from A. Adrian Albert, February 9, 1933, University of Göttingen archives.
2. J.J. O'Connor and E.F. Robertson, http://www-groups.dcs.st-and.ac.uk/~history/Mathematicians/Artin.html.
3. Constance Reid, *Hilbert* (New York: Springer-Verlag, 1996), at p. 205.
4. Saunders Mac Lane "Mathematics at the University of Göttingen 1931–1933," in J. Brewer and M. Smith, eds., *Emmy Noether: A Tribute to Her Life and Work* (New York: Marcel Dekker, Inc.), 1981, at p. 69.
5. J.J. O'Connor and E.F. Robertson, "John von Neumann," http://www-gap.dcs.st-and.ac.uk/; Albert Messiah, *Quantum Mechanics* (Mineola, N.Y.: Dover Publications, 1999), at pp. 190-191; John von Neumann, *Die Mathematische Grundlagen der Quantenmechanik* (1932).
6. Nick Herbert, *Quantum Reality: Beyond the New Physics* (New York: Anchor Books, 1985), at p. 25.
7. Bruce Schechter, *My Brain is Open* (New York: Simon and Schuster, 1998); Fred Jerome, *The Einstein File*, (New York: St. Martin's Griffin, 2003); Academy of Achievement, "Edward Teller, Ph.D., Father of the Hydrogen Bomb," http://www.achievement.org/; Gene Dannen, "Leo Szilard—A Biographical Chronology," http://www.dannen.com/chronbio.html.
8. Letter to Mrs. Adrian Albert from Eleanor Blumenthal, November 19, 1935, Albert family papers.
9. A Purdue University "Memorial Resolution" honoring Schilling reads: "In the fall of 1939 he went as an instructor to the University of Chicago through the sponsorship of another great algebraist, A.A. Albert." (Email to N.E. Albert from P. Roquette.)
10. Letters to Mrs. A.A. Albert from Dorothy Stanton (Schilling), November 29 and December 11, 1939, Albert family papers.
11. Letter to Frieda D. Albert from A. Adrian Albert, September 7, 1939, Albert family papers.

12. Arthur M. Schlessinger, Jr., ed., *The Almanac of American History* (New York: Barnes & Noble Books, 1993).
13. Fred Jerome, *The Einstein File* (New York: St. Martin's Griffin, 2002), at pp. 30-31.
14. Letter to Franklin D. Roosevelt from Albert Einstein, August 2, 1939, quoted in Daniel Boorstin, ed., *An American Primer* (Chicago: University of Chicago Press,1966), at pp. 857-862; Laura Fermi, *Illustrious Immigrants* (Chicago: University of Chicago Press, 1968), at p. 184.
15. Laura Fermi, *ibid.*, at p. 184; John Newhouse, *War and Peace in the Nuclear Age* (New York: Alfred A. Knopf, 1989), at p. 20.
16. Letter to Frieda from Adrian Albert, September 7, 1939, Albert family papers.
17. Daniel Zelinsky, "A.A. Albert," *Amer. Math. Monthly* vol. 80, no. 6 (June-July 1973), 662, reprinted in R. Block et al., eds., *A. Adrian Albert, Collected Mathematical Papers*, (2 vol's.), (Providence, R.I.: American Mathematical Society, 1993), at p. xl.
18. *Chicago Daily Tribune*, March 21, 1941.
19. Interview with Robert Oehmke and Theresa Oehmke, April 16, 2003.
20. Professional Biography of A.A. Albert, reprinted in R. Block et al., eds., *A. Adrian Albert Collected Mathematical Papers* (2 vols.) (Providence, R.I.: American Mathematical Society, 1993), at p. xxxiii.
21. *Webster's 3rd New International Dictionary*, unabr., (Springfield, Mass.: G. & C. Merriam, 1976); David Kahn, *The Codebreakers* (New York: MacMillan, 1967), at pp. xiii-xvi.
22. Letter to Mrs. A.A. Albert from Eleanor Blumenthal, September 17, 1941, Albert family papers.

Harvard University's Prof. George D. Birkhoff and his son, Prof. Garrett Birkhoff, also attended the dinner party. George Birkhoff wrote afterwards, "Dear Albert: May I express deep appreciation of both Mrs. Birkhoff and myself for the most delightful evening with you and Mrs. Albert on Friday last. It was an especial pleasure to us both to learn to know Mrs. Albert. It was a great privilege to be with you and your friends, and we will remember the occasion. You and the others responsible certainly arranged the whole mathematical occasion in a most admirable way." (Letter to A.A. Albert from George D. Birkhoff, September 7, 1941, Albert family papers.)

23. Kahn, *supra*, note 21, at p. 405.

24. A. Adrian Albert, "Some Mathematical Aspects of Cryptography," in *A. A. Albert Collected Mathematical Papers*, vol. 2, *supra*, note 20, 903-920, at p. 903. In a letter written to Adrian prior to publishing his scholarly history of cryptology, David Kahn asked whether Hill's work "may have stimulated other mathematicians to look at cryptology mathematically and so to apply mathematical tools" to cryptology. (Letter to Dr. A. Adrian Albert from David Kahn, March 12, 1965, Albert family papers.) In his response, Adrian acknowledged that Hill's effort was in some sense "pioneering," but countered, "The real power of the algebraic method in mathematics was developed after Hill's contribution to cryptography...." (Letter to David Kahn from A.A. Albert, March 24, 1965, Albert family papers.)

25. Kahn, *supra*, note 21, at pp. 410, 737-739.

26. A. Adrian Albert, "Some Mathematical Aspects of Cryptography," *supra*, note 24, at p. 903. (Emphasis in original.)

27. Kahn, *supra*, note 21, at p. 410.

28. "Math Students Help Create Secret Codes," *The Daily Maroon*, Thursday, January 9, 1941.

29. Letter to A.A. Albert from Neil Grabois, Williams College, Williamstown, Massachusetts, November 4, 1969; letter to Neil Grabois from A.A. Albert, November 10, 1969, Albert family papers.

30. Letter to Dean Tomlinson Fort from A.A. Albert, January 12, 1942, AMS archives, Box 28, Folder 45.

31. Http://www4.nationalacademies.org.

32. "Manhattan Project," http://college.hmco.com/history.

33. Letter to Mr. and Mrs. Adrian Albert from Mayme Logsdon, Brea College, Brea, Kentucky, September 10, 1944, Albert family papers.

34. A.A. Albert, "Address to 1951 Ray School Graduating Class," unpub., Albert family papers.

35. A.A. Albert, typewritten notes, undated, University of Chicago archives.

36. A.A. Albert address to 1951 class, *supra*, note 34.

37. A.A. Albert notes, *supra*, note 35.

38. A. A. Albert notes, *supra*, note 35.

39. A.A. Albert began working as a Research Mathematician for the Applied Mathematics Group at Northwestern University in September 1944. He became Associate Director in January 1945 and served in that capacity until the project ended in November 1945. (See Professional Biography in *A. Adrian Albert Collected Mathematical Papers*.) When the project started up, Prof. E.J. Moulton was named as Northwestern

University's Technical Representative under the government contract, while Prof. Walter Leighton, Jr. was designated as Director of Research. In January 1945, Prof. Moulton was transferred from the project to full-time teaching at Northwestern, and Leighton assumed the dual role of Technical Representative and Director of Research. (Letter to Dr. Warren Weaver, The Rockefeller Foundation, from Franklyn B. Snyder, President, Northwestern University, December 21, 1944, Northwestern University Archives, Franklyn Bliss Snyder Papers, Series 3/16/1, Box 39, Folder 17.)

40. Interview with E. Kleinfeld.

41. Letters to A.A. Albert from Saunders Mac Lane, August 24, 1944 and August 29, 1944.

42. Interview of Leon Cohen by William Aspray and Albert Tucker, Princeton Oral History Project, "The Princeton Mathematics Community in the 1930s," April 13, 1984, Tr. 6.

43. Email to N.E. Albert from I. Kaplansky, March 29, 2004; J.J. O'Connor and E.F. Robertson, "Irving Kaplansky," http://www-history.mcs.st-andrews.ac.uk/history.

44. Irving Kaplansky, "Abraham Adrian Albert," National Academy of Sciences, *Biographical Memoirs* (1980), 3-22, at p. 8; email to N.E. Albert from I. Kaplansky, January 29, 2003.

45. Saunders Mac Lane quoting Magnus Hestenes, "The Applied Mathematics Group at Columbia in World War II," in *A Century of Mathematics in America*, Peter Duren et al., eds., vol. 3 (Providence, R.I.: American Mathematical Society, 1988), 495, at p. 511.

46. Letter from to Frieda from Adrian Albert, September 13, 1943, Albert family papers.

47. Letter to Mrs. A.A. Albert from Tech. Sgt. Albert Davis, February 21, 1944, Albert family papers.

48. Letter to Prof. A.A. Albert from Gilbert A. Bliss, Flossmoor, Illinois, December 22, 1944, Albert family papers.

49. Letter to "Dearest" from Adrian, September 20, 1943, Albert family papers.

50. Letter to Adrian and Frieda Albert from David Cohodes, August 8, 1944, Albert family papers.

51. Letter to Mrs. A.A. Albert from Tech. Sgt. Albert Davis, September 13, 1944, Albert family papers.

52. Saunders Mac Lane, "Mathematics at the University of Chicago: A Brief History," in Peter Duren et al., eds., *A Century of Mathematics in*

America, vol. 2 (Providence, R.I.: American Mathematical Society, 1988), 127-154.

53. "The Reader's Companion to American History," http://college.hmco.com/history/readerscomp/rcah/rcag/html/ah_056000_manhattanpro.htm.

54. Jack M. Holl, *Argonne National Laboratory 1946–96* (Urbana: University of Illinois Press, 1997), at p. 7.

55. McGeorge Bundy, *Danger and Survival: Choices about the Bomb in the First Fifty Years* (New York: Random House, 1988), at pp. 18-23.

56. "Werner Heisenberg (1901–1976)," http://www.malaspina.com/site/person_623.asp.

57. Bundy, *supra*, note 55, at p. 20; John Newhouse, *War and Peace in the Nuclear Age* (New York: Alfred A. Knopf, 1989), at p. 10.

58. Holl, *supra*, note 54, at p. 8.

59. Fermi, *supra*, note 14, at p. 190.

60. Donald Fleming editorial in *An American Primer*, *supra*, note 14, at p. 862; Fermi, *supra*, note 14, at p. 199.

61. Newhouse, *supra*, note 57, at p. 42.

62. *New York Times*, March 15, 1958, quoted in Fred Jerome, *supra*, note 13, at p. 61.

63. Letter to S. Lefschetz from A.A. Albert, March 27, 1946, University of Chicago archives.

14: South of the Border

1. Interview with Zelda Zelinsky, February, 2004.
2. Email to N.E. Albert from I. Kaplansky, March 29, 2004.
3. Interview with Paul Sally, Jr., March 24, 2004.
4. Letter to "Dear Frieda" from Ruth Billingsley, June 18, 1972.
5. Letter to S. Lefschetz from A.A. Albert, March 3, 1946, University of Chicago archives.
6. "100 Mathematicians Will Assemble Here," *The Princeton Herald*, December 13, 1946; Letter to Adrian Albert from Solomon Lefschetz, May 13, 1946, University of Chicago archives.
7. Letter to Solomon Lefschetz from A.A. Albert, May 16, 1946, University of Chicago archives.
8. "President Truman Leaves Today," *Brazil Herald*, "Brazil's Only English Language Daily," September 7, 1947; "Hyde Parkers, All!", *Hyde Park Herald*, Oct. 26, 1988, at page 31.

9. "Report to the Division of International Exchange Persons Submitted by A.A. Albert," June 23, 1947, University of Chicago archives.
10. *Ibid*.
11. *Brazil Herald*, June 11, 1947.

15: Hyde Park in the Forties and Beyond

1. Memorandum to the Council of the American Mathematical Society from A.A. Albert, Managing Editor, *Transactions*, December 4, 1946, AMS archives, Box 31, Folder 124.
2. *Ibid*.
3. Richard Bellman, *Eye of the Hurricane: an Autobiography* (Singapore: World Scientific, 1984).
4. Recapitulation of events in letter to A.A. Albert from R. Bellman and H.N. Shapiro, April 13, 1948, AMS archives, Box 34, Folder 12.
5. Bellman, *supra*, note 3.
6. Letter to Dr. R. Bellman and Dr. H.N. Shapiro from A.A. Albert (quoting original retraction letter), April 20, 1948, AMS archives, Box 34, Folder 12.
7. *Ibid*.
8. *Ibid*.
9. Letter to J.R. Kline, Secretary, A.M.S., from A.A. Albert, AMS archives, Box 34, Folder 12.
10. Letter to A.A. Albert from J.R. Kline, Secretary, A.M.S., April 2, 1948, AMS archives, Box 34, Folder 12.
11. *Supra*, note 4.
12. *Supra*, note 6.
13. Bellman, *supra*, note 3.
14. Letter to A.A. Albert from Solomon Lefschetz, April 22, 1948, AMS archives, Box 34, Folder 12.
15. Letter to Dr. R. Bellman and Dr. H.N. Shapiro from A.A. Albert, April 24, 1948, AMS archives, Box 34, Folder 12.
16. Letter to A. Adrian Albert from Leonard Blumenthal, University of Missouri, Columbia, Missouri, December 1, 1949, Albert family papers.
17. Letter to "Dear Frieda" from Ruth Boorstin, 8 June 1972, Albert family papers.

18. A.A. Albert, "Address to 1951 Ray School Graduating Class," unpub., 1951, Albert family papers.
19. Ibid.
20. Robert M. Hutchins, quoted in Carroll Mason Russell, *The University of Chicago and Me: 1901–1962* (Chicago: Carroll Mason Russell, 1982), at p. 48.
21. Robert M. Hutchins, *The University of Utopia* (Chicago: University of Chicago Press, 1953), at p. 45.

16: Albert Algebras

1. A. Adrian Albert, "A Note on the Exceptional Jordan Algebra," *Proc. Nat. Acad. Sci.*, vol. 36, no. 7 (1950), 372-374; A. Adrian Albert. "On A Certain Algebra of Quantum Mechanics," *Ann. of Math.*, vol. 35, no. 1 (1934), 65-73.
2. A. A. Albert, "A Construction of Exceptional Jordan Division Algebras," *Ann. of Math.*, vol. 67, no. 1 (1958), 1-28, at p. 1; H. P. Petersson, "Albert Algebras: Their Meaning for Present-day Mathematics," unpub. email to N.E. Albert from Holger P. Petersson, June 10, 2003.
3. Nathan Jacobson, "Abraham Adrian Albert 1905–1972," *Bull. Amer. Math. Soc.*, vol. 80 (1974), 1075-1100, at p. 1088; Petersson, *ibid*.
4. See, *e.g.*, C. Chevalley and R. Schafer, "The Exceptional Simple Lie Algebras F_4 and E_6," *Proc. Nat. Acad. Sci.*, vol. 36 (1950), 137-141; H. Freudenthal, "Beziehungen der E_7 und E_8 zur Oktavenebene," I-XI, *Indag. Math.* 1954–1963.
5. Petersson, *supra*, note 2.
6. A. Adrian Albert, "A Construction of Exceptional Jordan Division Algebras, *Ann. of Math.*, vol. 67, no. 1 (1958), 1-28; Interview with R. B. Brown, April 21, 2003; Petersson, *supra*, note 2.
7. Jacobson, *supra*, note 3, at p. 1088.
8. Email to N.E. Albert from H.P. Petersson, June 26, 2003.
9. *Ibid*.
10. A.A. Albert and N. Jacobson, "On Reduced Exceptional Simple Jordan Algebras," *Ann. Math.*, vol. 66, no. 3 (1957), 400-417; Jacobson, *supra*, note 3, at p. 1088.
11. A.A. Albert and L. J. Paige, "On A Homomorphism Property of Certain Jordan Algebras, *Trans. Amer. Math. Soc.*, vol. 93 (1959), 421-439.
12. Email to N.E. Albert from J.M. Osborn, June 23, 2003.

13. A.A. Albert, "A Solvable Exceptional Jordan Algebra," *J. Math. Mech.*, vol. 8 (1959), 331-337.
14. A.A. Albert, "On Exceptional Jordan Division Algebras," *Pacific J. Math.*, vol. 15, no. 2 (1965), 377-404.
15. Kevin McCrimmon, "The Freudenthal-Springer-Tits' Constructions of Exceptional Jordan Algebras," *Trans. Amer. Math. Soc.*, vol. 139 (1969), 495-510.
16. Jacobson, *supra*, note 3, at p. 1088.
17. Petersson, *supra*, note 8.
18. Email to N.E. Albert from Louis Rowen, May 1, 2003; Petersson, *supra*, note 8.
19. Some of the other mathematicians who have conducted research on *Albert algebras* are: Jacques Tits, H. Freudenthal, M. Rost, J.P. Serre, E. Zelmanov, T.A. Springer, Alberto Elduque, Hyo Myung, Kevin McCrimmon, R. Parimala, Maneesh Thaker, and Alexandrov Iltyakov.
20. Email to N.E. Albert from H.P. Petersson, May 14, 2003; H.P. Petersson and M. Racine, "Enumeration and classification of Albert algebras: reduced models and the invariants mod 2," ed. S. Gonzalez, *Non-associative Algebra and Its Applications*, 334-340 (1994), Kluwer Academic Publishers, Netherlands.
21. Kevin McCrimmon, *A Taste of Jordan Algebras* (New York: Springer-Verlag, 2003).
22. J.J. O'Connor and E.F. Robertson, "John von Neumann," http://www-groups.dcs.st-andrews.ac.uk/.
23. Petersson, *supra*, note 2.

17: Defending His Country

1. Letters to A.A. Albert from Detlev Bronk, National Research Council, September 20, 1949 and April 21, 1969, University of Chicago archives.
2. "Appoint U. of C. Teacher Head of Math Group," *Chicago Tribune*, August 3, 1953; Letter to "Dear Colleagues" from Lipman Bers, Chairman, Division of Mathematical Sciences, National Academy of Sciences, May 21, 1970, University of Chicago archives.
3. Fred Kaplan, *The Wizards of Armageddon* (New York: Simon and Schuster, 1983), at p. 39.

4. A.A. Albert: (1) "A Two-Plane Campaign against Two Classes of Targets," RAND, December 3, 1951; (2) "A Two-Plane Attack against Three or Four Targets," RAND, February 16, 1953.

5. A.A. Albert, *Fundamental Concepts of Higher Algebra* (Chicago: University of Chicago Press, 1956), at p. v.

6. Interview with J. Marshall Osborn, April 6, 2003.

7. Letter to A.A. Albert from Pres. Detlev Bronk, National Academy of Sciences, July 17, 1952, University of Chicago archives.

8. Http://www.nas.edu/about/.

9. Interview with Erwin Kleinfeld, April 23, 2003.

10. David Kahn, *The Codebreakers* (New York: Macmillan, 1967), at pp. 674-675.

11. Certificate of Appreciation presented to A. Adrian Albert by Vice Admiral Noel Gayler, U.S. Navy, Albert family papers.

12. Letter to Dr. Merle Andrew from A.A. Albert, April 13, 1964, University of Chicago archives.

13. Letter to Chancellor L.A. Kimpton from A.A. Albert, February 18, 1959, University of Chicago archives; Irving Kaplansky, "Abraham Adrian Albert, A Biographical Memoir," *Biographical Memoirs*, National Academy of Sciences, vol. 51 (1980), 3-32, reprinted in R. Block et al., eds., *A. Adrian Albert Collected Mathematical Papers* (Providence, R.I.: American Mathematical Society, 1993).

14. I. N. Herstein, "A. Adrian Albert," *Scripta Mathematica*, vol. 29, nos. 3-4 (1973).

15. Email to N.E. Albert from I. Kaplansky, January 3, 2003; interview with Eugene Paige, August 28, 2004.

16. Jack M. Holl, *Argonne National Laboratory 1946–96* (Urbana: University of Illinois Press, 1997), at p. x.

17. Steven Levy, *Crypto* (New York: Penguin, 2001), at p. 40.

18. Interview with Richard Schafer, January 10, 2003.

19. Daniel Gorenstein, "The Classification of Finite Simple Groups, A Personal Journey: The Early Years," in *A Century of Mathematics in America*, Peter Duren et al., eds., vol. 1 (Providence, R.I.: American Mathematical Society, 1988), at p. 448.

20. Letter to Neil R. Grabois, Department of Mathematics, Williams College, Williamstown, Massachusetts, from A.A. Albert, November 10, 1969, Albert family papers.

21. Letter to "Dear Children," from Frieda D. Albert, June 12, 1957, Albert family papers.

22. Http://www.nsf.gov/pubs/stis1994/nsf8816.txt.

23. Letter to Prof. Lowell Paige, Chairman, Department of Mathematics, University of California at Los Angeles, from A. A. Albert, October 14, 1965.

24. Green later wrote: "Recently, I was reminiscing about people I had worked for, and keep coming back again and again to the reflection that Adrian was the most satisfactory person I ever knew to work for or with. He was a fine man, by no means perfect—he was too human for that—but one that I and many others will always remember with affection and admiration:" Green's letter recalled "the hard work we did together when I was leg man for the 'Albert Survey'...." (Letter to "Dear Frieda" from John Green, 28 November 1972, Albert family papers.)

25. *The Chicago Maroon*, November 3, 1967, at p. 2.

26. Letter to Mr. James E. Cross, Secretary, Institute for Defense Analyses, from George W. Beadle, April 14, 1967, University of Chicago archives.

27. *The Codebreakers, supra*, note 10, at p. 677.

28. An IDA official later reported that Adrian's directorship of IDA's Communications Research Division had been "critical to its survival and subsequent vigor...[and his] crucial role in starting this effort...has been of extraordinary value to U.S. Security." (Letter to Mrs. A. Adrian Albert from Lee P. Neuwirth, September 29, 1989, Albert family papers.)

29. Interview with Reuben Sandler; interview with Eugene Paige.

30. Press Release, University of Chicago Office of Public Relations, June 6, 1962, University of Chicago archives.

31. JPL Purchase Requisition, May 17, 1961, Albert family papers.

32. Letter to Dr. A. Adrian Albert from Marshall S. Carter, Lieutenant General, U.S. Army, Director, National Security Agency, July 28, 1969, University of Chicago archives.

18: The University's Solution to Urban Decay

1. Rossi and Dentler, *The Politics of Urban Renewal* (New York: The Free Press of Glencos, 1961), Ch. 2.

2. Muriel Beadle, *The Hyde Park-Kenwood Urban Renewal Years, a History to Date* (Chicago: Muriel Beadle, 1964).

3. N.E. Albert, A. Beamon, P. Beck, H. Dow, and J. Kelman, "University Apartments," unpub. paper, 1971.

4. *Ibid.*

19: East of the Tracks

1. Jean Block, *Hyde Park Houses, An Informal History* (Chicago: University of Chicago Press, 1978), at p. 80 and plate 51; "East Hyde Park," *Hyde Park Herald*, Summer, 1979, at p. 13.
2. Conversation with Debra S. Goldberg, January 2002.
3. Irving Kaplansky, *Selected Papers and Other Writings* (New York: Springer-Verlag, 1995), at p. 65.
4. Ethel May Kharasch, "How Adrian Forgot His Axioms," January 1956.

20: Years of Triumph and Tragedy

1. Interview with Harley Flanders, who served as A.A. Albert's Research Assistant in 1946–47 while a student at Chicago.
2. Letter to S. Lefschetz from A.A. Albert, May 16, 1946, University of Chicago archives.
3. Marshall H. Stone, "Reminiscences of Mathematics at Chicago," *University of Chicago Magazine*, Autumn 1976, at p. 30.
4. Felix Browder, "The Stone Age of Mathematics on the Midway," *University of Chicago Magazine*, Autumn 1976, at p. 28; Saunders Mac Lane *et al.*, "The University of Chicago Department of Mathematics: A Brief History of the Department," www.math.uchicago.edu/history.html, at p. 2.
5. Browder, *ibid.*, at p. 28.
6. "Appoint U. of C. Teacher Head of Math Group," *Chicago Tribune*, August 3, 1952.
7. Nathan Jacobson, *Bull. Amer. Math. Soc.*, vol. 80 (1974), 1075-1100 at pp. 1077-1078.
8. Letter to Vice-President H. W. Harrison, University of Chicago, from A. A. Albert, January 2, 1954, University of Chicago archives.
9. A.A. Albert, "Report of Current Activities," December, 1954 (unpub.), University of Chicago archives.
10. Interview with R. Schafer, February 13, 2003; interview with J.M. Osborn, April 6, 2003; interview with R. Block, April 19, 2003.
11. Interview with J. M. Osborn, April 2, 2003.
12. Interview with Paul Sally, Jr., March 24, 2004. Sally's observation was based on seeing lecture notes taken by Adrian's students during the late 1950s.

13. Interview with Richard Block.
14. Interviews with Benjamin Muckenhoupt, December 31, 2003 and January 18, 2004.
15. Noether's protégé, Olga Taussky, wrote, "I absolutely adore" this theorem, although "it took quite a while before I had an application for it." (Olga Taussky, "My Personal Recollections of Emmy Noether," in James W. Brewer and Martha K. Smith, eds., *Emmy Noether: A Tribute to Her Life and Work* (New York: Marcel Dekker, 1981), at p. 92.)
16. A.A. Albert and Benjamin Muckenhoupt, "On Matrices of Trace Zero," *Michigan Math. J.*, vol. 4 (1957), 1-3.
17. Ben H. Yandell, *The Honors Class* (Natick, Mass.: A.K. Peters, 2002), at p. 73; interview with Eugene C. Paige, August 28, 2004.
18. "Report of Current Activities of A. A. Albert," *supra*, n. 9.
19. Saunders Mac Lane, "The Education of Ph.D.s in Mathematics," Peter Duren et al., eds., *A Century of Mathematics in America*, vol. 3 (Providence, R.I.: American Mathematical Society, 1988), 517-523.
20. *Connecticut Daily Campus*, November 29, 1956.
21. A.A. Albert, "On Certain Trinomial Equations in Finite Fields," *Ann. of Math.*, vol. 66 (1957), 170-178.
22. J.J. O'Connor and E.F. Robertson, "Nathan Jacobson," http://www-groups.dcs.st-andrews.ad.uk/.
23. A.A. Albert, "A Construction of Exceptional Jordan Division Algebras, *Ann. Math.*, vol. 67, no. 1 (1958), 1-28, at p. 28 (emphasis added).
24. A.A. Albert and N. Jacobson, "On Reduced Exceptional Simple Jordan Algebras," *Ann. Math.*, vol. 66, no. 3 (1957), 400-417.
25. D. Gorenstein, "The Classification of Finite Simple Groups, A Personal Journey: The Early Years," Peter Duren et al., eds., *A Century of Mathematics in America*, vol. 1 (Providence, R.I.: American Mathematical Society, 1988) 447-476, at p. 448.
26. Interview with Erwin Kleinfeld, April 23, 2003.
27. Gorenstein, *supra*, note 25.
28. Gorenstein, *supra*, note 25, at p. 448.
29. Interview with R. Schafer, January 10, 2003.
30. Email to N.E. Albert from Irving Kaplansky, January 8, 2003.
31. Letter to Prof. A.J. Lohwater, Chairman, Department of Mathematics, Case Western Reserve University, from A.A. Albert, Dean, February 12, 1969, University of Chicago archives.
32. Conversation with Roy Albert, November 9, 2002.

33. Letter to Prof. Deane Montgomery from A.A. Albert, December 1, 1958, Institute for Advanced Study archives.

34. Conversation with Prof. Robert Soare, October 18, 2002.

35. A.A. Albert and John Thompson, "Two Element Generation of the Projective Unimodular Group," *Illinois Journal of Mathematics*, vol. 3, no. 3 (September 1959), 421-439; W. M. Kantor, "Finite Geometry for a Generation," http://www.ulb.ac.be/assoc/bms/Bulletin/bul943/KANTOR.F.

36. A.A. Albert, "Two Element Generation of a Unimodular Group," unpub., January 9, 1957.

37. Interview with Erwin Kleinfeld, April 23, 2003; interview with Eugene C. Paige, August 28, 2004.

38. "John Griggs Thompson," J.J. O'Connor and E.F. Robertson, http://www-groups.dcs.st-andrews.ac.uk/.

39. Letter to Chancellor L.A. Kimpton from A.A. Albert, February 18, 1959, University of Chicago archives.

40. Letter to Prof. Angus Taylor, Chairman, Department of Mathematics, UCLA from Adrian Albert, April 13, 1959, University of Chicago archives; "$30,000 award to Albert," *Chicago Maroon*, April 3, 1959; "Prof. Albert gets $30,000 Math Grant," *Hyde Park Herald*, April 1, 1959.

41. A. Adrian Albert, "On the Collineation Groups Associated with Twisted Fields," *Calcutta Math. Soc., Golden Jubilee Commemoration Volume*, 1958/59, Part II. Calcutta Math. Soc., Calcutta, 1963, 485-497.

42. Letter to Chancellor L.A. Kimpton from A.A. Albert, February 18, 1959, University of Chicago archives.

43. Daniel Gorenstein, "On the Structure of Certain Solvable Groups," *Proceedings of a Symposium in Pure Mathematics of the American Mathematical Society*, (Providence, R.I.: American Mathematical Society, 1959), at p. 15.

44. A.A. Albert and Irving Kaplansky, eds., *Proceedings of a Symposium in Pure Mathematics of the American Mathematical Society* (Providence, R.I.: American Mathematical Society, 1959).

45. A.A. Albert and L.J. Paige, "On a Homomorphism Property of Certain Jordan Algebras," *Trans. Amer. Math. Soc.*, vol. 93 (1959), 20-29.

46. Irving Kaplansky, "Abraham Adrian Albert, A Biographical Memoir," *Biographical Memoirs*, (National Academy of Sciences, vol. 51, 1980), at pp. 9-10.

47. D. Gorenstein, *A Century of Mathematics in America, supra*, note 25, at p. 449.

48. Letter of reference from A.A. Albert, July 20, 1965, University of Chicago archives.

49. Adrian referred to his success in obtaining formal authorization to hire Nash in a letter to Professor L.J. Paige, Chairman, Department of Mathematics, UCLA, March 28, 1959. (Albert family papers).

50. Sylvia Nasar, *A Beautiful Mind* (New York: Simon & Schuster, 1998), at p. 244.

51. *Ibid.*, at p. 244.

52. Marshall H. Stone, *supra*, note 3, at p. 31.

53. Letter to Dear Colleagues from A.A. Albert, June 7, 1961, University of Chicago archives.

54. *Ibid.* Two years hence, in 1960, A.A. Albert finally succeeded in generating an offer for Jacobson to teach at Chicago. But the offer came too late. After so many false starts, Jacobson elected to remain at Yale. He later described some of his reasoning for declining the position in his *Collected Papers*.

55. Letter to Dear Colleagues, *id.*

21: Urban Removal and Condomania

1. N.E. Albert, A. Beamon, P. Beck, H. Dow, and J. Kelman, "University Apartments," unpub. (1971), at p. 7.

2. *Ibid.*, at p. 10.

22: E. H. Moore at the Entry

1. Letter to Mr. Adrian A. Albert [sic] from Lawrence A. Kimpton, July 11, 1960, Albert family papers.

2. A.A. Albert, undated typewritten notes, University of Chicago archives.

3. Letter to Prof. H.D. White from G. Bliss, January 23, 1933, University of Chicago archives; Karen H. Parshall, "Eliakim Hastings Moore and the Founding of a Mathematical Community in America, 1892–1902," Peter Duren et al., ed., vol. 2, *A Century of Mathematics in America* (Providence, R.I.: American Mathematical Society, 1988), at p. 165.

4. AMS website, ams@ams.org; J.J. O'Connor and E.F. Robertson, http://www-groups.dcs.st-andrews.ac.uk/.

5. Letter to G. Bliss from Prof. H.S. White, January 27, 1933, University of Chicago archives.

6. Letter to Mrs. Moore from George D. Birkhoff, January 1, 1933, University of Chicago archives.

7. L.E. Dickson, quoted in Parshall, *supra*, note 3, at p. 175.

8. *Hyde Park Herald*, Wednesday, July 6, 1960.

9. Handwritten note from the desk of Harry S. Everett, Albert family papers.

10. Letter to "Dear Adrian" from Saunders Mac Lane, July 20, 1960, Albert family papers.

11. Letter to Professor A. Adrian Albert from Daniel J. Boorstin, Professor of American History, July 22, 1960, Albert family papers.

12. Letter to Professor Adrian Albert from Alan Simpson, Dean of the College, July 13, 1960, Albert family papers.

13. Letter to "Dear Adrian" from Willis B. Foster, Executive Secretary, Advisory Panel on General Sciences, Office of the Director of Defense Research and Engineering, July 15, 1960, Albert family papers.

14. A. Adrian Albert and Reuben Sandler, *An Introduction to Finite Projective Planes* (New York: Holt, Rinehart and Winston1968), at p. v.

15. *See, e.g.*, A.A. Albert: (1) "On the Collineation Groups Associated with Twisted Fields," *The Golden Jubilee Commemoration Volume (1958–1959)*, Calcutta Mathematical Society; (2) "On the Collineation Groups of Certain Non-Desarguesian Planes," *Separata de Portugaliae Mathematica*, vol. 18 (1959), 207-224.

16. A.A. Albert, "Isotopy for Generalized Twisted Fields," *An. Acad. Brazil. Ci.*, (1961), 265-275.

17. Irving Kaplansky, "Abraham Adrian Albert," (*Biographical Memoirs*, Nat. Acad. Sci., vol. 51, 1980), 3-22 at p. 9.

18. A.A. Albert, "The Finite Planes of Ostrom," *Bol. Sci. Mat. Mexicana* (2) 11 (1966), 1-13; T.G. Ostrom, "Semi-translation Planes," *Trans. Amer. Math. Soc.*, vol. 111 (1964), 1-18.

19. A.A. Albert, "Finite Planes for the High School," *The Mathematics Teacher*, vol. LV, no. 3 (1962).

20. A. Adrian Albert, *Solid Analytic Geometry* (Chicago: Phoenix Books, University of Chicago Press, 1966).

21. A.A. Albert, "Finite Division Algebras and Finite Planes," *Proc. Sympos. Appl. Math.*, vol. 10, (Providence, R.I.: Amer. Math. Soc., 1960), 53-70.

22. Albert and Sandler, *An Introduction to Finite Projective Planes*, *supra*, note 14.

23. *See, e.g.*, Block, Jacobson, Osborn, Saltman, and Zelinsky, eds., (2 vols) *A. Adrian Albert, Collected Mathematical Papers* (Providence, R.I.: American Mathematical Society, 1993).
24. Letter to A. Adrian Albert from John J. Courtney, Jr., Lt. Col., U.S. Air Force, Deputy Chief, Office of Research and Development, National Security Agency, November 4, 1954.
25. A.A. Albert, "On Certain Polynomial Systems," *Scripta Math.*, vol. 28 (1967), 15-19.
26. Conversation with Izaak Wirszup, October 24, 2002.
27. Interview with A.A. Albert in "A History of IDA's Association with the University," *The Chicago Maroon*, November 3, 1967, p. 2.
28. Confidential source.
29. Interview with Reuben Sandler, January 4, 2003; interview with Richard Block, April 19, 2003.
30. A. Adrian Albert response to questionnaire for National Register of Scientific and Technical Personnel, Mathematics and Statistics, American Mathematical Society, April 2, 1964.
31. Nathan Jacobson, "Abraham Adrian Albert, 1905–1972," *Bull. Amer. Math. Soc.*, vol. 80 (1974), 1075-1100, at p. 1093.

23: An Entirely New Dimension

1. Emails from Dean David Oxtoby and Prof. Stephen Stigler to N.E. Albert, October 28, 2002.
2. Letter to "Dear Adrian" from Marshall Stone, December 27, 1961, Albert family papers.
3. Letter to Prof. A. Adrian Albert from Harold R. Ellman, Vice-President, Hart Schaffner & Marx, January 5, 1962, Albert family papers.
4. Letter to Prof. A. Adrian Albert from Julia Wells Bower, Connecticut College, January 8, 1962, Albert family papers.
5. Letter to "Dear—ugh—Dean Albert" from H.S. Everett, January 8, 1962, Albert family papers.
6. Letter to Prof. A.A. Albert from William H. Meyer, January 23, 1962, Albert family papers.
7. Letter to "Dear Adrian" from Leonard M. Blumenthal, University of Missouri Department of Mathematics, March 26, 1962, Albert family papers.

8. A.A. Albert, "On Exceptional Jordan Division Algebras," *Pacific Journal of Mathematics,* vol. 15, no. 2 (1965), 377-404.

9. Interview with David Saltman, September 30, 2003; A.A. Albert: (1) "New Results on Associative Division Algebras," *Journal of Algebra,* vol. 5, no. 1 (1967), 110-132; (2) "On Associative Division Algebras," *Bull. Amer. Math. Soc.,* vol. 74 (1968), 438-454; (3) "A Note on Certain Cyclic Algebras," *Journal of Algebra,* vol. 14 (1970), 70-72.

10. J. Minac and A. Wadsworth, "The First Two Cohomology Groups of Some Galois Groups," http://www.math.ucsd.edu/wadsworth.MW.html.

11. A.A. Albert, ed., *Studies in Modern Algebra,* vol. 2 (Englewood Cliffs, N. J.: Mathematical Association of America, 1963) (Prentice-Hall, distrib.). Prof. George Glauberman describes the book as "an excellent collection of articles, successfully designed to be accessible to graduate students and advanced undergraduates. It is also an excellent reference for a class on nonassociative algebras...." (Letter to N.E. Albert from George Glauberman, September 21, 2003.)

12. Interview with Reuben Sandler, December 21, 2003.

13. A. Adrian Albert and Reuben Sandler, *An Introduction to Finite Projective Planes* (New York: Holt, Rinehart and Winston,1968), at p. 18.

14. Letter to Dr. Milton Rose, National Science Foundation, from A. Adrian Albert, December 7, 1964, University of Chicago archives.

15. Dean A. Adrian Albert, "The Next Ten Years of the Division of the Physical Sciences," unpub., 1965, University of Chicago archives.

16. "U. C.'s Freedom Spurs Science Research," *Chicago Tribune,* November 25, 1963.

17. *Ibid.*

18. *Id.*

19. A.A. Albert, "Address to the 50th Annual Meeting of the National Association of Plant Administrators of Universities and Colleges," delivered at the University of Chicago's Center for Continuing Education May 20, 1963, unpub., University of Chicago archives.

20. "5 1/2 Million U. of C. Chemistry Plan Told," *Chicago Tribune,* October 2, 1963.

21. *Ibid.*

22. "New Computer Center at U. C. Gets $500,000," *Chicago Tribune,* February 10, 1963.

23. Handwritten letter to A.A. Albert from Marshall H. Stone, September 30, 1965, University of Chicago archives.

24. A.A. Albert, University of Chicago Convocation Address, "Federal Support of the Physical Sciences in Universities," unpub., 1971, University of Chicago archives.

25. "City's Science Role Boosted During Year," *Chicago Tribune*, January 1, 1964.

26. Ioan James, *Remarkable Mathematicians* (Cambridge, England: Cambridge University Press and Mathematical Association of America), 2002; Theoni Pappas, *Mathematical Scandals* (San Carlos, CA.: Wide World Publishing/Tetra), 2002.

27. Conversation with Sol K. Newman, June, 1998.

28. "Nobel Prize in Chemistry Won By U. Of C. Researcher," *Chicago Sun-Times*, November 4, 1966, at p. 4.

29. Letter to Professor Adrian Albert, Dean, from Solomon Golomb, December 16, 1963, University of Chicago archives.

30. Letter to Prof. Zohrab Kaprielian, Chairman, Department of Electrical Engineering, University of Southern California, from A. A. Albert, January 6, 1964, University of Chicago archives.

31. Steven Levy, *Crypto: How the Rebels Beat the Government—Saving Privacy in the Digital Age* (New York: Viking Penguin, 2001), at pp. 39-41.

32. "Albert Elected by Math Society," *Hyde Park Herald*, January 1, 1964; "U. C. Dean Wins Mathematical Society Honor," *Chicago Tribune*, January 2, 1964.

24: Turmoil on Campus

1. Seymour Hersh, *The Dark Side of Camelot* (Boston: Little, Brown, 1997), at p. 344.

2. "Governor George Wallace," http://www.stanford.edu/tommyz/1960's/George%20Wallace.htm.

3. "Integration Now!" www.teacher.scholastic.com/researchtools/articlearchives/.

4. Lyndon B. Johnson, *My Hope for America* (New York: Random House, 1964), at p. 32.

5. *Ibid.*, at p. 51.

6. Hersh, *supra*, note 1, at p. 412.

7. Hersh, *supra*, note 1, at p. 413.

8. *The Chicago Maroon*, November 3, 1967, at p. 1.

9. *Ibid.*, at p. 2.

10. "'Think Tank' Put on the Defensive," *Washington Post*, March 30, 1968.
11. Letter to Prof. A.J. Lohwater, Case Western Reserve University, from A.A. Albert, February 12, 1969, University of Chicago archives.
12. Interview with R.B. Brown, April 21, 2003.
13. Http://speccoll.library.kent.edu/.
14. *Ibid.*

25: Education Guru

1. A. Adrian Albert, Convocation Address, University of Chicago, "The Goal of Excellence in a University," August 30, 1963, unpub., University of Chicago archives.
2. A. Adrian Albert, Convocation Address, *ibid.*; "Avoid Mediocrity, U.C. Grads Urged," *Chicago Daily News*, August 30, 1963.
3. A.A. Albert, "The Curriculum in Mathematics," unpub., November, 1964, University of Chicago archives.
4. Letter to Dean A. Adrian Albert from Governor Otto Kerner, July 17, 1964, University of Chicago archives.
5. A.A. Albert, "Review: A Program on Mathematics, Systems Research, and Engineering," unpub., January 29, 1965, University of Chicago archives.
6. Letter to A.A. Albert from Theodore Ashford, Assoc. Dean and Dir., Natural Sciences and Mathematics Division, University of South Florida, April 4, 1969, University of Chicago archives.
7. Letter to Dean A. Adrian Albert from Prof. Arnold Ross, December 21, 1970, University of Chicago archives.
8. Letter to Dr. James A. Robinson, Vice President for Academic Affairs and Provost, Ohio State University, from A. A. Albert, April 16, 1971, University of Chicago archives.
9. Congratulatory letter to A.A. Albert from Gordon Walker, May 22, 1968; letter from A.A. Albert to Alberto Gonzalez-Dominguez, June 7, 1968, University of Chicago archives.
10. Press release, Institute for Advanced Study, August 18, 1969.
11. Letter to Dr. Carl Kaysen, Director, Institute for Advanced Study, from A. Adrian Albert, August 11, 1969, University of Chicago archives.
12. Letter to Mr. Donald Straus from A. Weil, Institute for Advanced Study, November 30, 1970, University of Chicago archives.

13. Letter to Prof. Deane Montgomery, School of Mathematics, Institute for Advanced Study, from A.A. Albert, January 11, 1971, University of Chicago archives.

14. Letter to "Dear Frieda" from André Weil, June 29, 1972, Albert family papers.

26: Building Bridges across the Globe

1. Olli Lehto, "IMU—Past and Present," http://www.mathunion.org.

2. Letter to Marston Morse, Esq., International Congress of Mathematicians, Low Memorial Library, New York City, from George E. Roosevelt, AMS archives, Box 27, Folder 72. Roosevelt was a banker and cousin of both FDR and Theodore Roosevelt.

3. *Proceedings of the 1950 International Congress of Mathematicians* (2 vols.) (Providence, R.I.: American Mathematical Society, 1952).

4. *Ibid.*

5. Report on Current Activities of A. Adrian Albert, December 1954, University of Chicago archives.

6. *Proceedings of the 1954 International Congress of Mathematicians* (3 vols.) (Amsterdam: North-Holland Publishing Co., 1956).

7. *Proceedings of the 1958 International Congress of Mathematicians* (Cambridge: Cambridge University Press, 1960).

8. *Proceedings of the 1962 International Congress of Mathematicians* (Djursholm, Sweden: Almquist & Wiksells Uppsala, 1963).

9. "Chicago U. Dean Talks with Arturo Illia," *Latin American Times*, New York, September 15, 1965.

10. University of Chicago Office of Public Relations (10/11/65 draft), University of Chicago archives.

11. Letter to Prof. A. Adrian Albert, Dean, from Wolfgang Mechbach, September 24, 1965, University of Chicago archives.

12. "University Teachers Begin Leaving Argentina," *New York Times*, August 19, 1966.

13. Letter to Dean A.A. Albert from Alberto Gonzalez-Dominguez, Universidad de Buenos Aires, September 4, 1966, University of Chicago archives.

14. "Albert Heads Unit for Math Event," *Chicago Tribune*, August 13, 1967.

15. Letter to "Dear Frieda" from Otto Frostman, Djursholm, Sweden, February 2, 1972, Albert family papers.

16. Letter to Prof. Jean Leray, College de France, from A. Adrian Albert, April 29, 1970, University of Chicago archives.
17. Letter to Miriam Ringo from A. Adrian Albert, October 27, 1969, University of Chicago archives.
18. *L'Espoir*, Merdi ler Septembre 1970.
19. *Actes du Congrès International des Mathématicians 1/10/Septembre 1970* (Nice, France: Gauthier-Villars, 1971).
20. International Congress of Mathematicians, "Fields Medal Prize," http://www.icm2002.org.cn/general/prize/fmedal.htm.
21. Letter to A.A. Albert from Nicolae Atanasiu, First Secretary, Embassy of the Socialist Republic of Romania, August 14, 1969, University of Chicago archives.
22. Letter to Dean A.A. Albert from Robert M. Forcey, Section on USSR and Eastern Europe, National Academy of Sciences, November 17, 1970, University of Chicago archives.
23. Letter to H. Barnett from A.A. Albert, May 25, 1971, University of Chicago archives.
24. Letter to Dr. Geoffrey Keller, Dean, College of Mathematics and Physical Sciences, Ohio State University, from A.A. Albert, May 21, 1971, University of Chicago archives.
25. "The Copernican View and the Search for Extraterrestrial Intelligence" and "Extraterrestrial Intelligence: Possible Research Directions," *News Report*, National Academy of Sciences, October 1971, at pp. 1 and 4.
26. Letter to Dean A. Adrian Albert from G.K. Skrjabin, Acting Chief Scientific Secretary of the Presidium of the Academy of Sciences of the USSR, July 16, 1971, University of Chicago archives.
27. Letter to Dr. A. Adrian Albert from Noel Gayler, Vice Admiral, U.S. Navy, 13 July 1971, University of Chicago archives.
28. University of Chicago, Office of Public Relations Press Release, August 26, 1971, University of Chicago archives.
29. Letter to Carolyn, Dan, and Brad from Adrian Albert, September 30, 1971, University of Chicago archives.

27: Diplomacy for Mathematics

1. Nathan Jacobson, "Abraham Adrian Albert, 1905–1972," *Bull. Amer. Math. Soc.*, vol. 80 (1974), at p. 1078.
2. *Ibid.*, at p. 1077.

3. A.A. Albert, typewritten report to the University of Chicago administration, 1954, University of Chicago archives.

4. A.A. Albert, "The Needs of American Mathematics," unpub., undated address, University of Chicago archives. A similar address by A.A. Albert entitled "The Support of Mathematics in the National Science Foundation" is dated November 13, 1952.

5. Letter to Prof. Leonard Gillman from A.A. Albert, February 27, 1963, University of Chicago archives.

6. Letter to Dr. Merle Andrew, Director of Mathematical Sciences, Office of Scientific Research, Department of the Air Force, from A. Adrian Albert, April 13, 1964, University of Chicago archives.

7. Letter to Mrs. Frieda Albert from Reinhold Baer, Zürich, Switzerland, 18 July 1972, Albert family papers.

8. A.A. Albert, *College Algebra* (New York: McGraw-Hill, 1946), at p. v. The text was Adrian's maiden effort at replacing undergraduate texts then extant. The University of Chicago Press published a revised version of the college algebra text in 1963.

9. A.A. Albert, *College Algebra, ibid.*

10. Christmas letter from Elsie and Ed Begle, December, 1972, Albert family papers.

11. Note to Frieda Albert from Ed and Elsie Begle, December, 1972, Albert family papers.

12. A. Adrian Albert, Preface to *Modern Higher Algebra* (Chicago: University of Chicago, 1937), at p. vii.

13. A. Adrian Albert, Preface to *Introduction to Algebraic Theories* (Chicago: University of Chicago Press, 1941), at p. v.

14. A.A. Albert, *College Algebra* (Chicago: University of Chicago Press, Phoenix Science Series, 1963), Preface, at p. v.

15. *Ibid.*, Preface, at p. viii.

16. A. Adrian Albert, *Solid Analytical Geometry* (Chicago: University of Chicago Press, Phoenix Science Series, 1966), Preface, at p. v.

17. *Ibid.*, at p. vi.

18. Letter to Victor Klee, University of Michigan, from A. Adrian Albert, January 9, 1970, University of Chicago archives; interview with W. Reid, August 1, 2004; interview with N. Lebovitz, August 2, 2004.

19. A.A. Albert, "Mathematics As A Profession," October 20, 1960, unpub., University of Chicago archives.

20. *Ibid.*

21. Jacobson, *supra*, note 1, at p. 1077.

22. A.A. Albert, "Mathematics During the Past Decade," February 12, 1965, unpub., University of Chicago archives.
23. Letter to A. Adrian Albert from Milton Rose, head of Mathematical Sciences Section, National Science Foundation, February 19, 1965, University of Chicago archives.

28: Mathematical Children

1. Interview with J. Marshall Osborn, April 6, 2003; interview with Robert Brown, April 21, 2003.
2. Interview with M.S. Frank, March 30, 2003; M.S. Frank, "A New Class of Simple Lie Algebras," *Proc. Nat. Acad. of Sci.*, vol. 40, 1954, 713-718.
3. A.A. Albert and M.S. Frank, "Simple Lie Algebras of Characteristic p," *Univ. e Politec. Torino. Rend. Sem. Mat.*, vol. 14 (1954–55), 117-139.
4. Letter to Mrs. A.A. Albert from M. Frank, September 21, 1972, Albert family papers.
5. Letter to "Dear Frieda" from Tony Huston, June 13, 1972, Albert family papers.
6. *Proc. Amer. Math. Soc.*, vol. 4, 1953, 444-451.
7. Interview with Richard Block, April 19, 2003.
8. Melikian Gaik, "New Simple Lie Algebras: Melikian Algebras," http://www.cs.uwm.edu/.
9. Interview of William L. Duren, Nathan Jacobson, and Edward J. McShane, Princeton Oral History Project, Transcript 8.
10. Interview with J.M. Osborn, April 6, 2003. Another former student recalled, "He was always willing to lend me a helping hand when I needed it and was certainly generous in giving his time to the entire mathematical community." (Letter to Mrs. Adrian Albert from Robert Oehmke, June 16, 1972.) Yet another acknowledged Adrian as both teacher and friend, saying that he owed Adrian a debt "which one cannot repay...[except by trying] to benefit and be of help to other students who are coming along now." (Letter to Mrs. Albert from R. B. Brown, June 8, 1972.)
11. Letter to "Dear Adrian" from Carl Kaysen, Institute for Advanced Study, July 6, 1971, University of Chicago archives.
12. A. Adrian Albert, ed., *The Collected Papers of Leonard Eugene Dickson* (6 vols.) (Bronx, N. Y.: Chelsea Publishing, 1975), a p. v.

13. Interview with R.B. Brown, 2003.
14. Conversation with Sandra L. Goldberg, August 2003.
15. Wheaton College archives; Wheaton Alumni Magazine online, www.Wheaton.edu/alumni/magazine.
16. D. Zelinsky, "A.A. Albert," *Amer. Math. Monthly*, vol. 80, no. 6 (June-July 1973), at p. 665.
17. Izaak Wirszup, Memorial address, June 8, 1972.
18. Edward Levi, President of the University, summed up A.A. Albert's contribution as follows: "Adrian was one of the truly great men of our University. He worked hard and endlessly for the University. He gave the University his love, his ideals, his perception of quality, his insistence upon the highest standards. He was one of the world's greatest scholars. What could have been personal pride, fully justified, he turned into renewed and selfless efforts to help the institution which, in so many ways, represented his best efforts. He had a fierce pride in the University, and he made the University a much better place.

"I had the privilege of working with him for many years. I saw him solve all kinds of problems which arise in a University—many of them were problems which inevitably were vexing and annoying—the kind of problems from which a person of less stature and devotion to the University would have backed away, knowing that the personal cost to himself was so great.

"I will miss him so very much, as I have missed him during these past months. His influence was, of course, most immediate in the Division of the Physical Sciences, but it was enormous also throughout the University. But his influence is still with us. He will always be remembered, of course, as a great mathematician, but as long as this University survives, it will reflect his efforts, his standards and his personality. And for those of us who knew him, we will continue to hear his voice, as I do now, on so many countless matters where he placed the good of the University before everything else." (Letter to "Dear Frieda" from Edward H. Levi, June 8, 1972, Albert family papers.)
19. *Supra*, note 16.
20. Honorary Doctor of Laws Certificate awarded by the University of Notre Dame to A. Adrian Albert, May 15, 1965.
21. N. Jacobson, "Abraham Adrian Albert (1905–1972)," *IMU Bulletin*, No. 4 (December 1972), at p. 7.
22. S.A. Amitsur, "On Central Division Algebras," *Israel Journal of Mathematics*, vol. 12, No. 4, 1972.

23. Letter to Dr. Pinchas Rosen, Chairman of the Rothschild Prizes Organization, from A.A. Albert, March 27, 1970, University of Chicago archives.

24. The issue featured scientific papers authored by some of Adrian's colleagues and former doctoral students. Former students who contributed papers for publication in the memorial issue included Robert B. Brown, Nathan Divinski, Robert Oehmke, J. Marshall Osborn, Reuben Sandler, and Richard Schafer. There was also a paper by John Thompson, the young mathematician whose career he had fostered.

25. I.N. Herstein, "On a Theorem of Albert," *Scripta Mathematica*, vol. XXIX, No. 3-4 (1973), at p. 391.

26. I.N. Herstein, "A. Adrian Albert," *Scripta Mathematica, ibid.*, at p. lxxvi.

27. Letter to A.A. Albert from Alberto Gonzalez-Dominguez, University of Hawaii, 16-I-68, University of Chicago archives.

28. Letter to Mrs. A.A. Albert from Lou Auslander, Graduate School and University Center of the City University of New York, June 29, 1982. The letter was accompanied by a 1981 article by L. Auslander and R. Tolimieri entitled, "A Matrix Free Treatment of the Problem of Riemann Matrices," *Bull. Amer. Math. Soc.*, vol. 5, No. 3 (November 1981).

29. Block, R., Jacobson, N., Osborn, J.M., Saltman, D., and Zelinsky, D., eds., *A. Adrian Albert Collected Mathematical Papers* (2 vols.) (Providence, R.I.: American Mathematical Society, 1993).

30. D.P. Jacobs, S.V. Muddana, and A.J. Offutt, "A Computer Algebra System for Nonassociative Identities," (1990), http://www.cs.clemson.edu.

31. Email to N.E. Albert from Irvin R. Hentzel, April 23, 2003; I. Hentzel, D.P. Jacobs, and E. Kleinfeld, "A Case Study Using Albert (1992)," http://citeseer.nj.nec.com/hentzel92ring.html.

32. D.P. Jacobs, "The Albert Nonassociative Algebra System: A Progress Report," (1994), http://www.cs.clemson.edu.

Select Bibliography

Albert, A. Adrian. *College Algebra.* New York: McGraw-Hill, 1946 (revised and reprinted, Phoenix Science series, University of Chicago Press, 1963).
———. *Fundamental Concepts of Higher Algebra.* Chicago: University of Chicago Press, 1956 (reprinted, Passaic, New Jersey: Polygonal Publishing House, 1981.)
———. *Introduction to Algebraic Theories.* Chicago: University of Chicago Press, 1941.
———. *Modern Higher Algebra.* Chicago: University of Chicago Press, 1937.
———. *Structure of Algebras.* New York: American Mathematical Society, Colloquium Publications, vol. 24, 1939 (revised printing, Providence, R.I.: AMS, 1961).
———. *Solid Analytic Geometry.* New York: McGraw-Hill, 1949 (revised printing, Phoenix Science series, University of Chicago Press, 1966).
Albert, A. Adrian, ed. *Collected Mathematical Papers of Leonard Eugene Dickson,* (6 vols.). Bronx, N.Y.: Chelsea Publishing, 1975.
Albert, A. A., ed. *Studies in Modern Algebra,* vol. 2: Englewood Cliffs, N.J.: Mathematical Association of America, 1963. (Prentice-Hall, distrib.)
Albert, A. A. and Kaplansky, I., eds. *Proceedings of a Symposium in Pure Mathematics of the American Mathematical Society.* Providence, Rhode Island: American Mathematical Society, 1959.
Albert, A. Adrian and Sandler, Reuben. *An Introduction to Finite Projective Planes.* New York: Holt, Rinehart and Winston, 1968.
Albert, N.E., Beamon, A., Beck, J., Dow, H., and Kellman, K., "University Apartments," unpub., 1971.
Appelbaum, Stanley. *The Chicago World's Fair of 1893.* New York: Dover Publications, 1980.

Barrow-Green, June. *Poincaré and the Three Body Problem*. History of Mathematics, vol. 11. American Mathematical Society and London Mathematical Society, 1997.

Bashmakova, I. and Smirnova. G. Shneitnzer, A., trans. *The Beginnings and Evolution of Algebra*. Mathematical Association of America, 2000.

Beadle, Muriel. "The Hyde Park-Kenwood Urban Renewal Years." Chicago: Muriel Beadle, 1964.

Bell, E. T. *The Development of Mathematics*. New York: Dover, 1992.

———. *Men of Mathematics*. NewYork: Simon & Schuster, 1937, 1965.

Birkhoff, Garrett, and Mac Lane, Saunders. *A Survey of Modern Algebra*, 5th ed. Wellesley, Massachusetts: A. K. Peters, 1997.

Block, Jean F. *Hyde Park Houses*. Chicago: University of Chicago Press, 1978.

Block, R., Jacobson, N., Osborn, J. M., Saltman, D., and Zelinsky, D., eds. *A. Adrian Albert Collected Mathematical Papers*, (2 vols). Providence, Rhode Island: American Mathematical Society, 1993.

Boorstin, Daniel J., ed. *An American Primer*. Chicago: University of Chicago Press, 1966.

Boyer, Carl and Merzbach, Uta. *A History of Mathematics*. New York: Wiley & Sons, 1968.

Brewer, James W. and Smith, Martha K., eds. *Emmy Noether: A Tribute to Her Life and Work*. New York: Marcel Dekker, Inc., 1981.

Bronson, Richard. *Linear Algebra*. San Diego: Academic Press, 1995.

Bundy, McGeorge. *Danger and Survival: Choices about the Bomb in the First Fifty Years*. New York: Random House, 1988.

Clawson, Calvin C. *The Mathematical Traveler*. Cambridge, Massachusetts: Perseus Publishing, 2003.

Condit, Carl W. *The Chicago School of Architecture*. Chicago: University of Chicago Press, 1964.

Corry, Leo. *Modern Algebra and the Rise of Mathematical Structures*. Basel: Birkhauser Verlag, 1996.

Curtis, Charles W. *Pioneers of Representation Theory*. Providence, Rhode Island: American Mathematical Society, 1999.

Cutler, Irving. *The Jews of Chicago*. Urbana and Chicago: University of Illinois Press, 1996.

Dean, Richard A. *Elements of Abstract Algebra*. New York: Wiley & Sons, 1966.

Deskins, W. E. *Abstract Algebra*. New York: Dover Publications, 1964.

Dick, Auguste. Blocher, H. I., trans. *Emmy Noether 1882–1935*. Boston, Basil, Stuttgart: Birkhäuser, 1981.
Dickson, Leonard Eugene. *Algebras and Their Arithmetics*. New York: Dover, 1960 (repub., 1923 ed.).
Du Sautoy, Marcus. *The Music of the Primes*. New York: HarperCollins, 2003.
Duren, Peter, et al., eds. *A Century of Mathematics in America* (3 vols). Providence, Rhode Island: American Mathematical Society, 1988.
Eaton, Ralph, ed. *Descartes Selections*. New York: Scribner's Sons, 1955.
Encyclopaedia Judaica, vol. 6. Jerusalem: Keter Publishing House Jerusalem Ltd., 1972.
Fermi, Laura. *Illustrious Immigrants*. Chicago: University of Chicago Press, 1968.
Fleming, Walter and Varberg, Dale. *College Algebra*. Englewood Cliffs, New Jersey: Prentice-Hall, 1980.
Freedman, Chaim. *Eliyahu's Branches: The Descendants of the Vilna Gaon and his Family*. Teaneck, New Jersey: Avotaynu, Inc., 1997.
Ginzberg, Louis. *Students, Scholars, and Saints*. Jewish Publication Society Press, 1928.
Goodspeed, Thomas W. *A History of the University of Chicago*. Chicago: University of Chicago Press, 1916.
⸻. *The University of Chicago Biographical Sketches* (2 vols.) Chicago: University of Chicago Press, 2d. ed., 1924–25.
Grattan-Guiness, Ivor. *Norton History of the Mathematical Sciences*. New York: W. W. Norton, 1997.
Greene, Brian. *The Elegant Universe*. New York: Vintage Books, 1999.
Halberstam, David. *The Fifties*. New York: Fawcett Columbine, 1993.
Harper, William Rainey. "University and Democracy." Chicago: University of Chicago Press, 1970.
Hasse, Helmut. *Higher Algebra*, vol. 1, Linear Equations. Benac, Theodore, trans., New York: Frederick Ungar Publishing, 1954.
Hawking, Stephen, et al., *The Future of Spacetime*. New York: W. W. Norton, 2002.
Henderson, Harry. *Modern Mathematics*. New York: Facts on File, Inc., 1996.
Herbert, Nick. *Quantum Reality: Beyond the New Physics*. New York: Anchor Books, 1985.
Hersh, Seymour M. *The Dark Side of Camelot*. Boston: Little, Brown, 1997.
Herstein, I. N. *Topics in Algebra*, 2nd ed. New York: Wiley. 1975.

Hoffmann, Banesh. *About Vectors*. New York: Dover Publications, 1966.
Holl, Jack M. *Argonne National Laboratory 1946–96*. Urbana: University of Illinois, 1997.
Hooper, Alfred. *Makers of Mathematics*. New York: Random House, 1948.
Hutchins, Robert M. *The University of Utopia*. Chicago: University of Chicago Press, 1953.
Hyde Park Historical Society. "Hyde Park History," no. 2. Chicago: Hyde Park Historical Society, 1980.
"The Institute for Advanced Study: Some Introductory Information." Princeton, New Jersey, 1980.
Jacobson, Nathan. *Collected Mathematical Papers*. Boston: Birkhauser, 1989.
James, Ioan. *Remarkable Mathematicians: From Euler to von Neumann*. Cambridge, U.K.: Cambridge University Press and Mathematical Assn. Of America, 2002.
James & James. *Mathematics Dictionary*. New York: Chapman & Hall International Thomson Publishing, 1992.
Jerome, Fred. *The Einstein File*. New York: St. Martin's Griffin, 2003.
Johnson, Lyndon B. *My Hope for America*. New York: Random House, 1964.
Kahn, David. *The Codebreakers*. New York: Macmillan Company, 1967.
Kaku, Michio. *Hyperspace*. New York: Doubleday Anchor Books, 1994.
Kaplan, Fred. *The Wizards of Armageddon*. New York: Simon & Schuster, 1983.
Kaplansky, Irving. *Selected Papers and Other Writings*. New York: Springer-Verlag, 1995.
Karush, William. *Webster's New World Dictionary of Mathematics*. New York: Webster's New World, 1989.
Knus, Max-Albert, Merkurjev, Alexander, Rose, Marcus, and Tignol, Jean-Pierre. *The Book of Involutions*. Providence, R.I.: American Mathematical Association, Colloquium Publications, vol. 44, 1998.
Levy, Stephen. *Crypto*. New York: Penguin Books, 2001.
Lowe, David. *Lost Chicago*. New York: Wings Books, 1975.
McGraw-Hill. *Dictionary of Mathematics*. New York: McGraw-Hill Book Company, 1988.
McGraw-Hill. *Encyclopedia of Science and Technology*, vol. 1. New York: McGraw-Hill Book Company, 2002.
McGraw-Hill. *Modern Men of Science*. New York: McGraw-Hill Book Company, 1966.

Meir, Golda. *My Life*. New York: G. P. Putnam's Sons, 1975.
Messiah, Albert. *Quantum Mechanics*. Mineola, N.Y.: Dover Publications, 1999.
Motz, Lloyd, and Weaver, Jefferson Hane. *The Story of Mathematics*. New York: Avon Books, 1993.
Muir, Jane. *Of Men & Numbers*. New York: Dover Publications, 1996.
Nahin, Paul J. *An Imaginary Tale: The Story of the Square Root of Minus One*. Princeton, New Jersey: Princeton University Press, 1998.
Nasar, Sylvia. *A Beautiful Mind*. New York: Simon & Schuster, 1998.
Newhouse, John. *War and Peace in the Nuclear Age*. New York: Alfred A. Knopf, 1989.
Newman, James R. *The World of Mathematics*. Tempus Books of Microsoft Press, 1988.
Nicholson, W. Keith. *Introduction to Abstract Algebra*. New York: Wiley & Sons, 1999.
O'Hara, Frank. "The University of Chicago: An Official Guide." Chicago: University of Chicago Press, 1928.
Osen, Lynn M. *Women in Mathematics*. Cambridge, Massachusetts: MIT Press, 1974.
Pappas, Theoni. *Mathematical Scandals*. San Carlos, CA: Wide World Publishing/Tetra, 2002.
Parkman, Francis. *The Discovery of the Great West: LaSalle*. New York: Rinehart & Company, 1956.
Popp, Richard. "Presidents of the University of Chicago." Chicago: University of Chicago Library. 1992.
Reid, Constance. *Hilbert*. New York: Springer-Verlag, 1996.
———. *Courant*. New York: Springer-Verlag, 1996.
Rossi and Dentler. *The Politics of Urban Renewal*. New York: Free Press of Glencos, Inc., 1961.
Russell, Carroll Mason. "The University of Chicago and Me: 1901–1962." Chicago: Carroll Mason Russell. 1982.
Sabbagh, Karl. *The Riemann Hypothesis*. New York: Farrar, Straus and Giroux, 2002.
Schechter, Bruce. *My Brain is Open*. New York: Simon & Schuster, 1998.
Schlesinger, Arthur M., Jr., ed. *The Almanac of American History*. New York: Barnes & Noble Books, 1993.
Scott, J. F. *The Scientific Work of René Descartes*. New York, London: Garland Publishing, 1987.
Segal, Sanford L. *Mathematicians under the Nazis*. Princeton: Princeton University Press, 2003.

Smith, David E. and Latham, Marcia L. *Geometry of René Descartes*. Chicago, London: Open Court Publishing, 1925.
Storr, Richard. *Harper's University: The Beginnings*. Chicago: University of Chicago Press, 1966.
Travers, K., Dalton, L., and Brunner, V. *Using Algebra*. River Forest, Illinois: Laidlaw Bros., Doubleday, 1945.
Truman, Harry S. *Years of Trial and Hope 1946–1952*. Garden City, N. Y.: Doubleday, 1956.
University of Chicago. *Cap and Gown*, vol. 31. Chicago: University of Chicago, 1926.
Van der Waerden, B. L. *A History of Algebra*. Berlin: Springer-Verlag, 1985.
Weisstein, Eric. *CRC Concise Encylopedia of Mathematics*. 2nd ed., Boca Raton: Chapman & Hall/CRC, 2003.

Index

Academy of the Socialist Republic of Romania, 244
Adams, John, 232
Adler, Dankmar, 18
Adler, Mortimer, 190
Adrian Albert Memorial Lectures, 265
Air Force Cambridge Research Center, 166, 168-169, 187, 194, 198
Airmail, 88
Alabama National Guard, 224
Albert, Alan D. (son), 40, 45, 145, 157-158, 165, 181, 193, 197, 207, 239-240
Albert, Elias (father), 7-11, 34-35, 39-40, 257
Albert, Fannie (mother), 8-11, 36, 150
Albert, Frieda (wife), 1-2, 11-14, 34-37, 40-42, 45, 52, 104-105, 109, 111, 122, 126, 128, 130, 136-138, 143-149, 155-156, 164-165, 167, 169, 177-179, 187, 196-197, 199-200, 203, 205-207, 210, 215, 227, 234-239, 242, 245-246, 261, 265
Albert, Joy (brother), 11
Albert, Minnie (sister), 9-10
Albert, Nancy E. (daughter), 2-3, 50, 145-146, 155-158, 178-180, 190-192, 206-207, 210, 259, 265
Albert, Roy David (grandson), 197
Albert, Roy Davis (son), 50, 112, 145-146, 157-158, 179, 189, 197
Albert-Brauer-Hasse-Noether theorem, 85-97
Albert interactive computer system for nonassociative algebras, 265-266
Albert survey, 170, 187, 251
Alexander, James, 45
Algebra
    abstract (modern), 29-31, 46, 64-74, 108, 110, 253
    definition of algebra, 60, 63-64

linear, 29, 64, 70
Algebras
    Albert algebras, 116, 159-162, 194
    Albert-Frank-Shalev algebras, 256
    alternative algebras, 116, 119
    associative algebras, 29, 38, 51, 73, 79-82, 108, 117, 210, 222
    Cayley algebras, 70-71
    central division algebras, 80-83, 216
    cyclic algebras, 81-82, 84, 86, 97-98, 108, 160-161, 216
    definition of algebras, 60-61
    Dickson algebras, 81-82, 84
    division algebras, 80-83, 86-87, 98
    exceptional Jordan algebras, 159-162, 194
    Jordan algebras, 115-121, 159-162, 194, 210, 216, 239, 265
    Lie algebras, 51, 115-117, 120, 187, 242, 255-256, 258, 260, 265
    Lie-admissible algebras, 119
    matrix algebras, 89
    nilpotent algebras, 260
    nonassociative algebras, 47, 73, 113, 115-121, 210, 222, 257, 265-266
    noncyclic algebras, 83, 87, 98, 216
    p-algebras, 97-98, 216, 255-256
    power-associative algebras, 119, 238
American Academy of Arts and Sciences, 232
American Association for the Advancement of Science, 48, 208
American Association of University Professors, 190
American Baptist Education Society, 19
American Council on Education, 109, 231
American Mathematical Society, 36, 82-83, 106-107, 109-112, 116, 127, 129-130, 187, 190, 200, 208, 222, 242-243, 265
American Red Cross, 136
American School of Abstract Algebra, 74
Amitsur, S. A., 263
Analysis (mathematical), 26, 29-30, 186
Analytic geometry, 66, 211, 216

INDEX   329

Anderson, Marian, 127
Annals of Mathematics, 38, 47, 106, 116, 193
Anti-Semitism, 35-36, 40, 47, 50, 101-103, 125-126, 139, 143, 178, 202, 213, 223
Argentina, 145, 151, 240-241
Argentine Academy of Sciences, 240-241
Argentine Mathematical Society, 240-241
Argonne National Laboratory, 168, 209, 258
    Applied Mathematics Division, 168
Armour, Philip, 16
Armour and Company, 43
Artin, Emil, 87, 89, 97, 125, 145
Atomic bombs, 139-141, 168
Atomic Energy Act of 1946, 168
Auslander, Louis, 105, 264
Austro-Hungarian Empire, 11
Axioms/Laws/Postulates, 59, 62, 73
    associative law, 73, 79, 116-117, 119
    commutative law, 62, 70, 73
    distributive law, 73
Babylonians, 65
Baer, Reinhold, 144, 238, 248
Baez, Joan, 178
Baptist Union Theological Seminary, 17, 21
Bartky, Walter, 177, 185, 214-215
Baton Rouge, 145
Baxter Laboratories, 259
Begle, Edward, 249
Belafonte, Harry, 178
Bell, E. T., 31, 66, 68
Bellman, Richard, 10 n.14, 152-155
Berman, Shelly, 178
Birkhoff, Garrett, 130 n.22, 144, 186, 238
Birkhoff, George D., 74, 130 n.22, 186, 208-209
Blighted Areas Redevelopment Act, 174

Bliss, Gilbert, 25 n.1, 28-30, 41, 74, 129, 137, 185-186
Block, Richard E., 120, 239, 257-258
Blumenthal, Eleanor, 126, 130, 239
Blumenthal, Leonard, 126, 155, 215, 239
Bolshoi Ballet, 246
Bolza, Oskar, 26, 28-29
Bombelli, Rafael, 67
Boorstin, Daniel, 2, 155, 209
Boorstin, Ruth, 2, 155
Bowdoin College, 168, 194, 200, 248
Boyce, Fannie, 260
Brauer, Richard, 46-47, 50-52, 86, 88-89, 92-95, 107, 127, 144-145, 201,238, 243-244, 253, 255-256
Brauer, Ilse, 52
Brazil, 145-152, 240
Brazilian Academy of Sciences, 149, 240
Browder, Felix, 186, 233, 250
Brown, Robert B., 256, 260
Brown v. Board of Education, 223
Bruck, Richard, 120 n.34, 144, 258
Bryn Mawr College, 46
Buenos Aires, Argentina, 151
Bulletin of the American Mathematical Society, 91, 94-95, 110-111, 131, 153
Burnham, Daniel, 18, 23
Burroughs, John C., 16
Caesar, Sid, 165
Calculus of variations, 26, 29-30, 129
Calderon, Alberto, 215
Cardano, Girolamo, 67
Carnegie Foundation, 126
Cartan, Eli, 258
Cartesian plane, 61, 66-68
Cayley, Arthur, 70-71, 79-80
Cayley algebras, 70-71

CDMA (code division multiple access), 80
Chandrasekhar, Subrahmanyan, 250
Chauvenet Prize, 30
Chern, Shiing-Shen, 179, 186, 199
Chevalley, C., 238
Chicago, city of
    Chicago City Council, 175-176
    Chicago Sanitary District, 43
    churches and synagogues
        Haddon Temple, 35
        Hyde Park Baptist Church, 180
        Sinai Temple, 180
    cultural institutions
        Art Institute, 18, 156
        Auditorium Theatre, 18
        Shedd Aquarium, 43
    history and founding, 15-16, 18-19, 22-24
    neighborhoods and suburbs
        Evanston, 26, 131, 173
        Humboldt Park, 10-11, 13, 27, 35
        Hyde Park, 1-2, 18-20, 155-156, 173-177, 179-180, 193, 197, 199, 203-206, 218
        Kenwood, 174-175
        Lawndale, 10
        Morgan Park, 17
        Morton Grove, 173, 259
        Near West Side (Maxwell Street neighborhood), 9-10
    parks
        Jackson Park, 18, 20, 23, 179-180
        Midway Plaisance, 19-20, 23, 138, 156
        Washington Park, 18, 20, 156
        Wooded Island, 180
    private clubs
        Hyde Park YMCA, 180
        South Shore Country Club, 178

schools
- Hyde Park High, 180
- Theodore Herzl Public Grammar School, 10
- John Marshall High, 10, 27, 180
- William H. Ray Public Grammar School, 148, 156, 177
- Tuley High, 13

urban renewal, 173-176, 203-206, 218

Chicago Conference on Algebra
- 1938 Conference on Algebra, 50-52, 129
- 1941 Conference on Algebra, 52, 129-130
- 1946 Conference on Algebra, 144

Chicago Land Clearance Commission, 174, 203

Chicago Mathematics Congress, 25-26, 29, 236

Chicago School of Algebra, 74

Chicago Sun-Times, 221

Chicago Title and Trust Company, 43

Chicago Tribune, 128, 217

Chicago, University of, 2, 11, 16-33, 37, 40, 74, 137-140, 142, 144, 171, 185-188, 190-194, 207-209, 212-221, 226-227, 254-256, 258
- admission of women students, 17, 20
- buildings
  - Administration Building, 227
  - Bond Chapel, 191-192
  - Computation Center, 218
  - Eckhart Hall, 2, 42-43, 138, 141, 207
  - Green Hall, 191
  - Fermi Institute, 218, 240
  - Harper Library, 192
  - Jones Chemical Laboratory, 42, 138, 218
  - Kent Chemical Laboratory, 42, 218
  - Law School, 192
  - Oriental Institute, 41-42
  - Quadrangle Club, 41, 203, 227
  - Rockefeller Chapel, 2, 36-37

Rosenwald Hall, 42
Ryerson Physical Laboratory, 42-43
Swift Hall, 192
Yerkes Observatory, 23, 42
Buildings and Grounds Department, 143, 215
Chicago Maroon, 131, 226
Department of Mathematics, 25-33, 122, 132, 142, 144, 177, 185-188, 193, 195-196, 198-199, 201-202, 207-209, 212-215, 219, 250, 254-256, 258, 262, 265
    history and founding, 16-24
    leadership role in mathematics, 26, 29, 186, 201-202
    leadership role in modern algebra, 74
Mathematics Club, 121
Physical Sciences Division, 177, 193, 213-222, 259, 262
prohibitions against gender and religious bias, 20
role in urban renewal, 174-175
Stagg Field, 2, 138, 140
Chile, 214, 240
Civil Rights Act, 225
Clark University, 39
Class field theory, 47, 86, 97-98
Cobb, Henry Ives, 23
Cohen, Paul, 189
Cohodes, Esther (half-sister), 9-11
Colby College, 187
Cole, Frank Nelson, 107
Cole Prize, 107
College of New Jersey (*see* Princeton University)
Collins, Judy, 178
Colloquium lectures, 26, 109-110, 208
Colloquium Publications, 109-110, 112
Columbia University, 40, 87, 134-136, 139
    Division of War Research, 134-136
    Teachers College, 134
Columbus, Christopher, 23

Combinatorial analysis, 117, 189, 195, 210-211
Compton, Arthur, 138-139
Computers, 64, 165, 168, 216-218, 250-251
Condominiums, 204-205
Continental Illinois Bank and Trust Company, 43
Copacabana Beach, 146-147
Cornell, Paul, 18-19
Cornell University, 51, 238
Courant, Richard, 99
Cryptanalysis, 129, 167-168
Cryptography, 64, 129-131, 167-169, 194, 222
Cryptology, 129-131, 167-168, 170-171, 211
Cuba, 235
D-Day, 140
Darrow, Clarence, 23
Davenport, Harold, 90, 100
Davidovics, Mordecai, 11-12
Davidovics, Yitta, 12
Davis, Daniel (brother-in-law), 136
Davis, Ethel (sister-in-law), 136
Davis, Hank (Edward) (brother-in-law), 136-137
Davis, Samuel (father-in-law), 12-13, 197
Davis, Rebecca (mother-in-law), 12, 35, 197
Dedekind, Richard, 68-69, 85
Denison University, 17-18
Desargues, Girard, 67
Descartes, René, 65-67
Deutsche Mathematik, 76, 108, 132
Dewey, John, 23
Dickson, Leonard Eugene, 26-27, 29, 31-33, 36, 38, 40-41, 43, 46, 51, 72, 74-75, 79, 82-88, 91, 126, 186, 208-209, 222, 231, 260
Diem, Ngo Dinh, 225-226
Dieudonné, Jean, 237-238
Diophantus, 65

Douglas Aircraft Company, 164
Douglas, Jim, Jr., 243, 250
Douglas, Stephen A., 16-17
Downs, James, Jr., 175
Dreyfus, Alfred, 35-36
du Sable, Jean Baptiste Point, 15
Duren, Peter, 146
Duren, Sally, 146
Duren, W. L., Jr., 31-33, 146
Dynamics, 62, 231
Eckhart, Bernard Albert, 42 n.30, 42-43
Eilenberg, Samuel, 144
Einstein, Albert, 45, 76, 125, 127-128, 141, 145
Einstein, Elsa, 45
Elijah ben Solomon (Vilna Gaon), 7-8, 48
Ellis Island, 12
Emergency Committee for the Aid of Displaced German Scholars, 126
England, 8, 127, 138, 165, 246
    Queen Elizabeth, 127
    King George VI, 127
    Lord Halifax, 165
    sale of arms to, 127
    Queen Victoria, 8
Equations, 59, 63
Encyclopedia Britannica, 187
Erie Railroad Company, 43
Erdős, Paul, 152-153
Escola Americana, 147-148
ESSO Education Foundation, 199
Everett, Harry Scheidy, 177, 209, 215
Evers, Medgar, 224-225
Existence theorem, 94, 100
Federal Housing Administration (FHA), 176
Feistel, Horst, 169, 222

Feit, Walter, 200-201, 243
Fenster, Della, 74, 88 n.10
Fermat, Pierre, 66 n.15
Fermi, Enrico, 139-140, 168, 240
Fermi, Laura, 139-140
Field, Marshall, 16, 20-21
Fields (mathematical) and field theory, 29, 51, 59, 64, 81, 165, 168, 193, 211
Fields, J. D., 244
Fields Prize, 239, 243-244
Florida State University, 221
Ford Foundation, 170
Fort, Tomlinson, 110, 131, 153
France, 35-36, 65-68, 127, 241-243, 246
    leadership role in mathematics, 65-68
Frank, Marguerite Straus, 120, 253-256
Frank-Wolfe algorithm, 256
Frobenius, Ferdinand Georg, 83, 86, 198
Frolic Theater, 155
Functions and function theory, 28, 30, 51, 59
Fundamental theorem of arithmetic, 69
Gainov, A. T., 119
Galileo, 65
Galois, Evariste, 67-68
Gauss, Johann Karl Friedrich, 27, 68
Gelfand, Israel M., 242
General Sciences Panel, 167, 247
Geometry, 26, 29, 60, 211-212, 216, 247, 250
    algebraic geometry, 48, 51, 253
    analytic geometry, 66
    differential geometry, 26
    projective geometry, 81, 211, 216
Germany, 12, 74-76, 125-127, 138-140, 236-237
    economic history, 74-75
    leadership role in mathematics, 68-70, 74-76

rise of Adolph Hitler, 75-76, 100, 126-127
  state university system, 75-76, 139
Gerstenhaber, Murray, 188, 257
GI Bill, 173
Glauberman, George, 216 n.11, 243
Goldberg, Debra (granddaughter), 179
Goldberg, Sandra (granddaughter), 260
Gompers, Samuel, 23
Gonzalez-Dominguez, Alberto, 241, 264
Good Will Hunting, 188-189
Goodspeed, Thomas, 19-21
Gorenstein, Daniel, 86, 194-195, 200-201
Gothic architecture, 23-24, 177, 203
Graves, Lawrence, 133
Great Chicago Fire, 16, 18
Great Depression, 40, 128, 260
Great October Socialist Revolution, 165
Great Society, 225
Green, John W., 170, 170 n.24, 187
Greenberg's Bakery, 155
Groups and group theory, 26, 29, 59, 64, 67-68, 70, 86, 114, 118, 131, 168-169, 195, 198, 200, 242-244
  Lie groups, 115
  quasigroups, 118-119
Gulf of Tonkin incident, 226
Hall, Marshall, 200
Halmos, Paul, 179, 185-186, 199
Hamilton, William, 60-62, 68, 70, 72-73, 80-81
Hamiltonian dynamics, 62
Hardy, G.H., 122
Harper, William Rainey, 17-18, 21-22, 53, 192
Harris Trust and Savings Bank, 43
Harvard University, 52, 107, 129, 185-186, 208, 253-254, 256
  Benjamin Peirce fellowships, 135, 142

Hasse, Helmut, 46, 86-103, 95 n.39, 100 n.61, 108, 125

Head Start program, 232

Heisenberg, Werner, 139

Hensel, Kurt, 96

Hershey, Gen. Lewis B., 133-134, 137

Herstein, I. N., 44, 167, 194-195, 200, 202, 250, 264

Herzl, Theodore, 34-36, 240

Hestenes, Magnus, 136

Hilbert, David, 25, 69, 75, 79, 85-86, 114, 236-237

Hill, Lester, 130, 130 n.24

Hille, Carl, 238-239

Hiroshima, 141

Houdini, 62

Housing Act of 1949, 175

Housing Act of 1954, 176, 203

Housing and Home Finance Agency, 176

Howe, Julia Ward, 23

Huston, Antoinette Killen, 31, 49, 257

Huston, Ralph, 49

Hutchins, Robert Maynard, 157-158, 185, 190, 196

Hyde Park Herald, 209

Hyde Park-Kenwood Community Conference, 174

IBM, 218, 222, 250, 260

Ideals (mathematical), 69, 188

Idempotents, 119-120, 160

IFF (Identification, Friend or Foe), 166

Illia, Arturo, 241-242

Illinois Institute of Technology (IIT), 194, 257

Illinois State Housing Board, 175

Index reduction factor theorem, 89, 93-94

Institute for Advanced Study, 44-47, 52, 112-115, 125-126, 197, 201, 233-234, 238, 259
    faculty dissension, 233-234
    founding, 44

Institute for Defense Analyses (IDA), 170-171, 200, 212-213, 226-227, 239, 259
Instituto Brasil-Estados Unidos, 150
Integral domain, 64
International Congress of Mathematicians (ICM), 26, 128, 196, 236-244
International Mathematical Union (IMU), 236-245, 259, 263
Invariant theory, 25
Involution, 106
Iron Curtain, 238
Isotopy, 117, 211
Israel, state of, 35-36, 240
    Herzl's role in founding, 35-36
    Tomb of Herzl, 240
Jacobson, Florence Dorfman, 52, 194
Jacobson, Nathan, 47-48, 50-52, 91, 97, 106, 115, 118, 122, 129, 144, 161, 194, 202, 202 n.54, 238, 247, 258, 263, 265
Japan, 127, 129, 140
Johns Hopkins University, 48
Johnson, Dorothy, 260
Johnson, Lyndon Baines, 225-226
Johnson, Warren, 193, 215
Joliet, Louis, 15-16
Jordan, Pascual, 115-117, 139, 159
Josef, Franz, 11
Journal für Mathematik, 92 n.31, 94, 96
Kahn, David, 130 n.24, 131
Kaplansky, Irving, 32, 50-51, 82, 89, 116, 118, 120, 135, 142-143, 152-153, 167-168, 179, 194-195, 201, 259
Kaplansky, Rachelle, 179
Katzenbach, Nicholas deBelleville, 192-193, 224
Kennedy, John F., 220, 223-226, 235
Kennedy, Robert, 223
Kent State University, 228
Kerner, Otto, 230
Kimpton, Lawrence, 174, 186, 207

Kingston Trio, 178
Klein, Felix, 25-26
Kleinfeld, Erwin, 120, 120 n.34, 194
Kline, John R., 110, 153, 238-239
Kokoris, Louis, 194, 257
Korean War, 164-165
Ku Klux Klan, 193, 224
Landau, Edmund, 108
Laurentiev, Mikhail, 243
Lebovitz, Norman, 250
Lefschetz, Solomon L., 38-40, 51, 105, 111, 114, 129, 141, 144-145, 154-155, 185, 254, 261
Lehigh University, 110
Lehrer, Tom, 178-179
Leighton, Walter, Jr., 134 n.39
Levi, Edward, 174, 227, 261 n.18
Levi, Julian, 174-175
Lie, Sophus, 114-115
Liebniz, Gottfried Wilhelm, 58, 67
The Limeliters, 178
Lincoln, Abraham, 18, 132
Lithuania, 7-8
  history, 7-8
  Jewish culture, 7-8
  Russian hegemony, 8
  Vilna, 7
Local-Global Principle, 95
Logsdon, Mayme, 133
Long, Huey, 146
Loops, 119
Louisiana State Capitol, 146
Louisiana State University, 145-146
Mac Lane, Saunders, 51, 121, 134-135, 144, 186, 193, 198, 209, 233, 238-239
Manhattan Project, 2, 138-141, 168, 220

Marburg University, 88
Mardi Gras, 145
Marquette, Jacques, 16-17
Marshall Field Foundation, 175
Maschke, Heinrich, 26, 30
Mason-Dixon line, 17
Massachusetts Bay Colony, 232
Massachusetts Institute of Technology (MIT), 202, 232, 257
Mathematical Association of America, 28, 30
Matrices and matrix theory, 38, 64, 70, 73, 79-80, 188-189, 210-212
McCormick, Cyrus, 16
McCrimmon, Kevin, 162
McLaughlin's Pharmacy, 156
McShane, Edward, 37 n.9, 238-239
Meir, Golda, 35
Meltzer, Jack, 175
Mentschikoff, Soia, 192
Mesta, Perle, 196
Metropolitan Housing and Planning Council, 174
Mexican Mathematical Society, 240
Mexico, 240-241
Meyer, Herman, 215
Mills, W., 194
Milnor, John, 239
Missouri Compromise, 17
MITRE Corporation, 171
Mittag-Leffler, Gosta, 220
Montgomery, Deane, 234, 238-239
Montgomery Ward and Company, 43
Moore, Eliakim Hastings, 25-26, 28-31, 33, 74-75, 143, 186, 207-209, 222
Moore, R. L., 74
Morse, Marston, 113, 133
Moscow, 239, 245-246, 260
Moulton, E.J., 134 n.39

Muckenhoupt, Benjamin, 189
Mulliken, Robert S., 220-221
Mussolini, Benito, 47, 139
Nachbin, Leopoldo, 149
Nagasaki, 141
Nasar, Sylvia, 202
Nash, John, 201-202
National Academy of Sciences, 132, 163, 165-166, 195, 220, 238-239, 242, 244-245, 262
National Aeronautic and Space Administration (NASA), 171
National Bureau of Standards, 139
National Research Council, 36, 113, 163, 165-166, 170, 187, 219, 259
  Division of Mathematical Sciences, 165, 187, 259
National Science Foundation, 169, 187, 193, 199, 201, 218, 230-231, 240, 247-248, 251-252, 257, 259-260
  NSF Survey on Training in Mathematics, 169, 187, 251
National Security Agency, 166-167, 170, 172, 187, 212, 222, 244-245, 257
National University of LaPlata, 240
New Christy Minstrels, 178
New Orleans, 145
New York Mathematical Society, 208
New Yorker, 196
Newberry, Walter, 16
Nicholescu, Miron, 244
Nichols-May, 178
Nobel, Alfred, 220
Nobel Prize, 139, 145, 220-221, 250
Noether, Amalie ("Emmy"), 46, 85-88, 92-95, 125-126, 188, 254-256
North Korea, 164, 180
North Vietnam, 226
Northwestern University, 25, 133-134
Notre Dame University, 222, 233
Number theory, 27, 46, 81, 86

Numbers
- algebraic, 81
- complex, 59, 61, 68-69
- decimals, 58
- fractions, 58
- hypercomplex, 46, 62, 73
- ideal numbers, 69
- imaginary, 58-59, 67, 69
- integers, 69
- irrational, 58
- negative, 58
- ordered pairs, 61
- π, 58
- positive, 58
- prime, 68, 111-112
- quaternions, 51, 62
- rational, 58
- real, 58-59
- square roots, 58
- unending decimals, 58
- whole numbers, 27, 58
- zero, 58

O'Connor, J. J., 198
Oehmke, Robert, 231, 256
Office of Naval Research, 113, 163, 200
Office of Ordnance Research, 120, 211
Office of Price Administration, 137
Office of Scientific Research and Development (OSRD), 132-136, 138-141, 164, 168, 248
- Applied Mathematics Panel, 133-136
- Manhattan Project, 2, 138-141

Ogden, William Butler, 16
Ohio State University, 228, 231-232, 244-245, 260
Olenik, Olga, 235-236

Olmsted, Frederick Law, 23
Operations (mathematical), 59
   addition, 59
   division, 59
   multiplication, 59
   subtraction, 59
Oppenheimer, Robert, 113
Osborn, J. Marshall, 118, 187, 239, 257-258
Ostrom, T. G., 211
Paige, Eugene C., 257-258
Paige, Lowell, 161, 194, 200, 239
Paige, Peggy, 194
Pall, Gordon, 32
Palmer, Potter, 16
Parshall, Karen H., 74
Pearl Harbor, 130
Peirce, Benjamin, 119, 142
Peru, 240
Petersson, Holger, 160-161
Poincaré, Henri, 105
Princeton University, 30, 36-40, 50, 139, 170, 232
   College of New Jersey, 37
   Department of Mathematics Advisory Council, 232
   Fine Hall, 45-46, 232
   Nassau Hall, 36
   Palmer Laboratory, 37
   Princeton Mathematics Club, 38
Proceedings of the National Academy of Sciences, 90, 256
Project ALP, 167-168
Projective geometry, 211, 216
Projective planes, 210-211
Puerto Rico, 232, 241
Purdue University, 202, 231
Pythagoras, 57

Quantum mechanics, 47, 64, 67, 71, 80, 114-115, 118, 125, 139-141, 150, 159
Quonset huts, 138
Racine, Michael, 161
Radcliffe College, 253
Rademacher, Hans, 111-112
Radios, 13, 27
RAND Corporation, 164-166, 200, 212
Rees, Mina, 133, 163
Reid, William H., 250
Relativity, theory of general, 70
Rensselaer Polytechnical Institute, 257
Representation theory, 86
Rice Institute, 230
Rice University, 230
Riemann, Bernhard, 68-70, 104-106, 111-112, 264
Riemann hypothesis, 111-112
Riemann matrices, 46, 104-106, 112, 114, 118, 208, 212, 264
Riemann surfaces, 70
Rings, 59, 64, 69, 79, 85, 118, 212
 associative rings, 118
 power-associative rings, 118
 simple alternative rings, 118
Rio de Janeiro, 145-150, 152, 155
Robertson, E. F., 198
Rockefeller Foundation, 126
Rockefeller, John D., 19, 21-22
Romania, 244
Roosevelt, Eleanor, 127
Roosevelt, Franklin Delano, 127-128, 132
Root, John Wellborn, 18, 23
Roquette, Peter, 89-90, 95-98
Rose, Milton, 252
Rosenwald, Julius, 42
Ross, Arnold, 32, 231

Rosser, John, 238-239
Roy D. Albert Prize, 197
Russia and Russians (*also see* U.S.S.R.) 164, 166, 235, 242, 244-246, 261
Sachs, Alexander, 128
Sagen, Oswald Karl, 28
Sandia Corporation, 166
Sandler, Reuben, 197, 211, 216, 257
Scalars, 60, 79
Schafer, Richard, 160, 194, 257
Schilling, Otto, 125, 126 n.9, 144, 202
School Mathematics Study Group (SMSG), 199, 249
Schur, I., 89
Scotland, 239
Scripta Mathematica, 259, 264
Sears, Roebuck Company, 42
Segal, Irving, 144, 186
Seligman, George, 194
Sets (mathematical), 58-59, 68, 117
Shapiro, Harold N., 152-154
Shoda, K., 188
Siegel, Carl Ludwig, 100-101, 112, 264
Simpson, Alan, 209
Slaught, Herbert Ellsworth, 28
Smith, Kate, 127
Snyder, Franklyn Bliss,134 n.39
South East Chicago Commission, 174-175
South Korea, 164, 180-181
South Vietnam, 225-227
Southern California Applied Mathematics Project (SCAMP), 166-168, 200, 207, 212
Spanier, Edwin, 186, 199
Sputnik, 169, 230, 251, 258
St. Gaudens, Augustus, 23
St. Xavier High School, 178

St. Xavier College, 200
Stanford University, 248-249
Stevenson, Adlai, 235
Stone, Harlan Fiske, 185
Stone, Marshall, 101-102, 129, 144, 185-186, 202, 214, 219, 229, 238-239
Strategic Air Command, 164
String theory, 64
Students for a Democratic Society (SDS), 226-227
Sullivan, Louis, 18
Suzuki, Michio, 200
Sweden, 239, 241
Swedish Royal Academy of Science, 221, 265
Switzerland, 236, 244-245
Szilard, Leo, 126-128
Taliaferro, William, 209
Taussky, Olga, 188 n.15
Teichmüller, Oswald, 108
Television, 155, 165
Teller, Edward, 126
Theodore Herzl Hall, 34
Thompson, John, 197-201, 233, 239, 243-244, 252
Tits, Jacques, 161
Toronto, 236, 253
Transactions of the American Mathematical Society, 90, 111-112, 152-155, 208
Truman, Harry, 148, 166, 168
Tukey, J., 238
Tulane University, 145-146
Turner, Frederick Jackson, 23
United Nations International Court of Justice, 238
U. S. Air Force Strategic Air Command, 164-165
U. S. Army Office of Ordnance Research, 120
U. S. Atomic Energy Commission, 163
U. S. Defense Department, General Sciences Panel, 167, 247
U. S. Department of Transportation, 256

U. S. State Department International Exchange Program, 145, 237, 240
United States Selective Service, 133-134, 137
Université d'Ottawa, 244
University of Alabama, 193, 224
University of Berlin, 86
University of Brazil, 145, 149-150
University of Buenos Aires, 145, 151
University of California at Berkeley, 139, 199
University of California at Los Angeles (UCLA), 195, 200, 248
University of California at Riverside, 257
University of Connecticut, 194
University of Göttingen, 75-76, 85, 88, 99
University of Illinois at Chicago, 257
University of Iowa, 231, 257
University of Kansas, 39
University of Königsberg, 47, 88
University of Maryland, 231
University of Michigan, 199
University of Minnesota, 139
University of Moscow, 235
University of Nebraska, 39
University of North Carolina, 50, 52
University of Pennsylvania, 238
University of São Paulo, 150
University of South Florida, 229, 231
University of Southern California, 200
University of St. Andrews, Scotland, 198
University of Toronto, 50, 253
University of Wisconsin, 128, 190-191, 258-259
Uraguay, 151
Urban Community Conservation Act, 172
Urban renewal, 173-176
U.S.S.R. (Soviet Union), 127, 165-166, 168-169, 235, 242, 244-246

van der Rohe, Mies, 203
van der Waerden, B.L.,67-68
Vassar College, 21
Veblen, Oswald, 45, 74, 111, 134, 145
Vectors, 60, 73, 211
Vector spaces, 60, 66
Veterans Administration, 173
Viet Cong, 226
Viète, François, 65 n.9
Vietnam War, 225-227
von Neumann, John, 45, 47, 113, 115-116, 125-126, 129, 145, 162, 165, 238-239
Wallace, George, 193, 224
War on Poverty, 225
Washington Post, 227
Wayne, Anthony, 15
Weaver, Warren, 134 n.39
The Weavers, 178
Wedderburn, Joseph Maclagen, 36-38, 47, 72-73, 75, 79-81, 85-86, 88-91, 121, 145
Weil, André, 186, 234, 238
Wessel, Caspar, 61 n.4
Western Hemisphere Mutual Defense Pact, 148
Weyl, F. J., 238
Weyl. Hermann, 45-46, 75, 114-115, 125-126, 145, 159
Wheaton College, 260
White, H.S., 25 n.1
Wick, Warner, 214
Wigner, Eugene, 115-116, 126-127, 139-140, 220
Wilczynski, Ernest, 28
Wilks, S. S., 238
Wilson, R. L., 258
Wirszup, Izaak, 212, 262
Witt, Ernst, 256
World War I, 13, 30, 37
World War II, 84, 125, 129-141, 163, 166, 168, 173, 185, 190, 229, 248, 251, 262

World's Columbian Exposition, 22-24, 180, 236
Yale University, 18, 21, 25, 168, 193-194, 199, 238, 249, 265
Yerkes, Charles T., 23, 42
Yeshiva University, 232-233, 259
Yiddish, 149
Zariski, Oscar, 47-48, 107, 129, 144-145, 238, 253-254, 256
Zassenhaus, Hans, 120 n.34
Zelinsky, Daniel, 32, 83, 106, 108, 118, 121, 135, 142, 239, 257, 262-263
Zelinsky, Zelda, 142
Zelmanov, Efim, 161
Zentralblatt für Mathematik, 101-103
Zionism, 36
Zorn, Max, 120 n.34
Zygmund, Antoni, 186

Printed in the United States
68661LVS00003B/161

PALO ALTO COLLEGE LRC
1400 W VILLARET BLVD
SAN ANTONIO, TX 78224

PALO ALTO COLLEGE LRC

A3 and His Algebra h

36171001544185